Additive Manufacturing (AM) of Metallic Alloys

Additive Manufacturing (AM) of Metallic Alloys

Editor

Flaviana Calignano

MDPI • Basel • Beijing • Wuhan • Barcelona • Belgrade • Manchester • Tokyo • Cluj • Tianjin

Editor
Flaviana Calignano
Politecnico di Torino
Italy

Editorial Office
MDPI
St. Alban-Anlage 66
4052 Basel, Switzerland

This is a reprint of articles from the Special Issue published online in the open access journal *Crystals* (ISSN 2073-4352) (available at: https://www.mdpi.com/journal/crystals/special_issues/Metallic_Alloys).

For citation purposes, cite each article independently as indicated on the article page online and as indicated below:

LastName, A.A.; LastName, B.B.; LastName, C.C. Article Title. *Journal Name* **Year**, *Article Number*, Page Range.

ISBN 978-3-03943-140-3 (Hbk)
ISBN 978-3-03943-141-0 (PDF)

Contents

About the Editor

Flaviana Calignano is an assistant professor at the Department of Management and Production Engineering (DIGEP) of the Politecnico di Torino (Italy). She obtained her Ph.D. in Industrial Production Systems Engineering from the Politecnico di Torino (Italy) in 2008. From 2005 to 2011, she worked on virtual reality in the study of facial morphology, WEB-based interactive 3D models for design and training, virtual machines for the simulation of machining, and reverse engineering. From 2012, her research activity has been focused on designing complex parts to be fabricated through additive manufacturing (AM), on the optimization of the process parameters, and on the study and definition of design rules for AM. She is the author and/or co-author of 5 book chapters, 1 patent, 1 monograph, more than 100 publications in national and international journals, and more than 40 communications to national and international scientific conferences.

Preface to "Additive Manufacturing (AM) of Metallic Alloys"

Developed in the 1980s, additive manufacturing (AM), known as rapid prototyping, has already revolutionized the production of polymeric material components. New developments in AM technologies are providing industries with the ability to build structural components with a variety of metal alloys, ceramics and composite materials. The introduction of metal AM processes has revolutionized the production of metallic components in the industrial sectors, where complex geometries, organic shapes, tubular, hollow designs, and dense, lattice-filled structures play a decisive role. In AM, there is no correlation between complexity and cost. Sometimes, more complexity means lower costs: less material and no need for assembly. However, there are problems that limit the wider uptake and exploitation of metals in AM. These range from the lack of design and modeling skills and AM software, to the different properties that are obtained using the same technology but different machines, to the difficulty of perfectly simulating the processes, to the incomplete understanding of the causes of variation in the quality of the parts, and to the repeatability of the processes.

Flaviana Calignano
Editor

crystals

MDPI

Editorial

Additive Manufacturing (AM) of Metallic Alloys

Flaviana Calignano

Department of Management and Production Engineering (DIGEP)-Integrated Additive Manufacturing Center (IAM)-Politecnico di Torino, Corso Duca Degli Abruzzi, 24-10129 Torino, Italy; flaviana.calignano@polito.it

Received: 14 August 2020; Accepted: 14 August 2020; Published: 15 August 2020

The introduction of metal additive manufacturing (AM) processes in industrial sectors, such as the aerospace, automotive, defense, jewelry, medical and tool-making fields, has led to a significant reduction in waste material and in the lead times of the components, innovative designs with higher strength, lower weight and fewer potential failure points from joining features.

This Special Issue on "Additive Manufacturing (AM) of Metallic Alloys" contains a mixture of review articles and original contributions on some problems that limit the wider uptake and exploitation of metal AM.

The variation in the quality of the parts and to the repeatability of the laser powder bed fusion process (L-PBF) was reviewed by Yadav et al. [1]. Their review focuses on the types of process defects that can be monitored via process signatures captured by in situ sensing devices and recent advancements in the field of data analytics for easy and automated defect detection.

One of the main causes that can lead to the poor quality of the components produced or the non-completion of the construction of the components is to be identified in the residual stresses. The fast and accurate prediction of residual stress is of great importance in the aerospace, automotive, and medical industries. Mirkoohi et al. [2] proposed a fully coupled thermomechanical analytical model to predict residual stress in the L-PBF process. In this process, due to the high-temperature gradient, the part experiences a high amount of thermal stress which may exceed the yield strength of the material. The proposed model uses a moving point heat source approach, the Green's function of stresses, and the Johnson-Cook flow stress model in order to predict the temperature field, the thermal stress, and the yield surface of the part under repeated heating and cooling, respectively.

The residual stress and other physical quantities, such as molten pool morphology, temperature field, and distortions, were examined by Saadlaoui et al. [3] in order to analyze and better understand the laser welding and melting of a powder bed fusion process. The results of this study can be used to improve and validate the numerical simulations of these processes.

Macroscale modeling plays an essential role in simulating AM processes. However, they cannot predict local quality problems such as a lack of fusion or surface roughness due to the fact that they often require a lot of computation time to get a certain accuracy of the results. Galati et al. [4] proposed an accurate and simple three-dimensional (3D) model to estimate the potential faulty process conditions that may cause quality issues or even process failure during the electron beam melting (EBM) process. Heat transfer analysis accuracy is demonstrated with a more accurate literature model. A multilayer simulation validates the model capability in predicting the roughness of a manufactured Ti6Al4V sample.

The fracture behavior and mechanical properties of the support structures for the components produced with this titanium alloy by the L-PBF process were investigated by Weber et al. [5]. In addition to fixing a part onto the building platform and the use of supporting overhanging structures, the support structures are necessary for heat dissipation, the avoidance of residual stresses and the compensation of residual stress-induced warping. The importance of support structures is often underestimated. This study is therefore based on the analysis of the process parameters that allow having suitable support structures for the purposes for which they are used. The experiments revealed that the struts produced with support parameters had no significantly lower tensile strength than the comparative parts.

The importance of design does not only concern the support structures but also, and above all, the components themselves. In fact, one cannot think of producing a component designed for traditional technologies with additive ones. Each technological process has its own design rules that depend on the constraints of the process itself. Galati et al. [6] investigated the designing for additive manufacturing (DfAM) and designing for assembly (DfA), to increase the material efficiency of the components for high-precision applications. The new methodology proposed allows a considerable reduction in the weight of the components and a simultaneous improvement in performance, as well as a reduction in the number of parts of a component that lead it to be economically competitive on an industrial level.

Remaining in the field of design, for the fabrication of mesoscale metal parts, Liu et al. [7] proposed the novel design of a low-cost electrochemical additive manufacturing (ECAM) 3D printer based on a microfluidic system. A copper circuit was printed on a non-conductive substrate to demonstrate a possible application of the ECAM system in the fabrication of functional electronics.

This Special Issue also includes a study conducted by Zhao et al. [8] on the microstructures and mechanical properties of ausrolled nanobainite steel. Due to their excellent mechanical properties and extraordinarily slender ferrite plates, high-C carbide-free nanostructured bainite steels are widely used in railway frogs, bearings, and automobile sheets in many industries.

Conflicts of Interest: The author declares no conflict of interest.

References

1. Yadav, P.; Rigo, O.; Arvieu, C.; Le Guen, E.; Lacoste, E. In situ monitoring systems of the SLM process: On the need to develop machine learning models for data processing. *Crystals* **2020**, *10*, 524. [CrossRef]
2. Mirkoohi, E.; Tran, H.C.; Lo, Y.L.; Chang, Y.C.; Lin, H.Y.; Liang, S.Y. Analytical modeling of residual stress in laser powder bed fusion considering part's boundary condition. *Crystals* **2020**, *10*, 337. [CrossRef]
3. Saadlaoui, Y.; Sijobert, J.; Doubenskaia, M.; Bertrand, P.; Feulvarch, E.; Bergheau, J.M. Experimental study of thermomechanical processes: Laser welding and melting of a powder bed. *Crystals* **2020**, *10*, 246. [CrossRef]
4. Galati, M.; Di Mauro, O.; Iuliano, L. Finite element simulation of multilayer electron beam melting for the improvement of build quality. *Crystals* **2020**, *10*, 532. [CrossRef]
5. Weber, S.; Montero, J.; Petroll, C.; Schäfer, T.; Bleckmann, M.; Paetzold, K. The fracture behavior and mechanical properties of a support structure for additive manufacturing of Ti-6Al-4V. *Crystals* **2020**, *10*, 343. [CrossRef]
6. Galati, M.; Calignano, F.; Viccica, M.; Iuliano, L. Additive manufacturing redesigning of metallic parts for high precision machines. *Crystals* **2020**, *10*, 161. [CrossRef]
7. Liu, P.; Guo, Y.; Wu, Y.; Chen, J.; Yang, Y. A low-cost electrochemical metal 3d printer based on a microfluidic system for printing mesoscale objects. *Crystals* **2020**, *10*, 257. [CrossRef]
8. Zhao, J.; Liu, D.; Li, Y.; Yang, Y.; Wang, T.; Zhou, Q. Microstructure and mechanical properties of tempered ausrolled nanobainite steel. *Crystals* **2020**, *10*, 573. [CrossRef]

crystals

MDPI

Article

Additive Manufacturing Redesigning of Metallic Parts for High Precision Machines

Manuela Galati *, Flaviana Calignano, Marco Viccica and Luca Iuliano

Department of Management and Production Engineering (DIGEP)-Integrated Additive Manufacturing Center (IAM)-Politecnico di Torino, Corso Duca Degli Abruzzi, 24-10129 Torino, Italy; flaviana.calignano@polito.it (F.C.); marco.viccica@polito.it (M.V.); luca.iuliano@polito.it (L.I.)

* Correspondence: manuela.galati@polito.it; Tel.: +39-011-090-4569

Received: 3 February 2020; Accepted: 26 February 2020; Published: 1 March 2020

Abstract: The conventional approach to design and manufacturing often has geometries with an efficient material distribution. For the high-precision machines, that approach involves the design of heavy components that guarantees the stiffness requirements. However, the higher the weight of the part, the higher inertia it has. As a result, when the feed axes are accelerated, the inertial forces deform the machine components and the precision of the machine is reduced. This study investigated the designing for additive manufacturing (DfAM) and designing for assembly (DfA) to increase the material efficiency of components for high-precision applications. A new methodology which considered the design and manufacturing issues and machining as well is given. A comprehensive model for cost evaluation of the part is presented. The case study refers to the rails and the bracket that support and move the flying probe of a testing machine for micro-electromechanical systems (MEMS). The weight of the rails has been decreased by 32% and the components to be assembled have been reduced from 16 to 7. The optimized bracket is more than 50% stiffer than the original one, 10% lighter, and economically competitive.

Keywords: design for additive manufacturing (DfAM); displacements; laser powder bed fusion (L-PBF); manufacturing constraints; stiffness; costs

1. Introduction

In recent years, the development of additive manufacturing (AM) technologies has led to numerous opportunities to fill the gap between optimal design and product application. The advantages of AM over conventional subtractive or formative methods clearly emerge when considering the great design freedom that can be achieved [1–3]. AM technologies allow fully dense and near-net-shaped parts to be produced with complex structures made of excellent materials. Industrial applications can be found for metal components [4], for which traditional manufacturing processes are expensive or difficult to apply [5]. The geometries that result from such design techniques as topology optimization (TO) [6–10] are examples of such geometries [1]. The so-called design for AM (DfAM) is being explored to show the design opportunities that are enabled by the adoption of both TO and AM [11,12]. At the current state-of-the art, components for structural applications are redesigned to achieve both weight reduction and performance improvement. In these components, TO has been conducted so far by introducing a decrease in stiffness [13] or using more performing materials specifically developed for AM technologies [14–16]. Such solutions may be effective in several fields, but they may not be practicable when the choice of new material involves a large increase in the cost of a component, or a much lower component stiffness. The stiffness of the structural components (such as the measuring probe, brackets, and the rails that support the measuring probes and motors) plays a key role in the accuracy of the machine, especially for high precision applications [17]. The use of conventional design methods and manufacturing systems lead to heavy and large components as final output [18] to guarantee high rigidity and limited deformations and vibrations [19,20]. However, the increased component weight (inertial forces) involves higher bending forces which may cause larger measurement errors [17]. Additionally, the accumulation of mass forces implies the use of lower traverse speeds that decrease machine productivity [21]. The aim of this research has been to show how such critical issues can be overcome by exploiting the benefit of DfAM. The study focused on the redesigning of some component of a high precision machine that uses flying probes to test the boards of micro-electromechanical systems (MEMS). Considering that only small production lots are produced per year and each testing machine is highly customized according to the specific requirements of each customer, the redesign lends itself well to production through metal AM systems. Because of the requirements of high dimensional accuracy, the laser powder bed fusion (L-PBF) technique is considered [22]. A comprehensive approach was proposed in which material distribution optimization, design for L-PBF, and design for assembly (DfA) are considered. This approach aimed also to overcome the current limitation of the design techniques and building volume of the L-PBF systems. The feasibility of the components and the machining operations have been investigated, and a comprehensive and detailed cost model has been developed and applied.

2. Design for L-PBF

During an L-PBF process, a laser source fuses a region of a metallic powder bed according to the computer aided design (CAD). When the one layer has been completed, the building platform is lowered, and the next layer of powder is deposited on the previous one. The process is then repeated until the part is completely built. Production by L-PBF (black flow line in Figure 1) involves several steps: the cleaning of loose powder from the part and post-process operations, including stress-relieving, removal of the supports, shot peening, heat treatments, and finishing operations [23].

The design of a part and its orientation on the building platform together with the choice of process parameters play a key role in the success of the process. From this point of view and according to one of the main rules for correctly using AM technologies, the design for L-PBF should focus on using the material only where necessary. Designing for L-PBF means considering during the design phases not only the constraints of the process, such as the minimum dimension of the feature, but also all activities that aim to guarantee the process and the part compliances, including the next manufacturing steps (support removal and machining). Here, the dimensional and the surface qualities and the metallurgic properties to be achieved on the part need to be considered [24].

For the part design, the DfAM and the DfA are the only possible design methodologies related to AM [25]. However, a comprehensive approach considers five steps: (1) the acquisition of the CAD model; (2) the part optimization; (3) the resulting new design; (4) the optimization of the orientation (purple square), and (5) the design verification. The optimized part from a structural point of view is the input for the subsequent steps. Advanced design techniques, such as TO, can be applied in this phase [2]. The optimized geometry needs to be checked under the geometrical limitation of the process. Adam et al. [26] provided a classification of these limitations, but it should be noted that design rules for L-PBF must not be considered as constraints but as modifiers during the design optimization [27]. Points 3 and 4 are iterative steps which consider reducing as little as possible the modifications of the optimize geometry resulting from step 2. Point 4, together with the support optimization, considers all of the above-mentioned activities. The evaluation of the optimal orientation is a hot topic in literature and usually is mainly based on the avoiding of support structures [28]. Leary et al. [29] proposed a methodology to evaluate the optimal build orientation according to the manufacturing time and component mass. However, because the L-PBF components are only near-net-shape, finishing operations should be carefully considered. Additionally, since support structures cannot be completely avoided, proper support design should be considered as leverage for the process [30].

The optimization phases (orientation and supports) are developed here with the aim to limit the support structure and the allowance for the machining operations. The iterative flow (red flow line in

3. Economic Analysis

The overall objective of the economic analysis is to estimate the manufacturing costs. Differently to the literature on the conventional manufacturing processes, the costs for metal AM processes have only been dealt with in a limited number of case studies, and a well-structured approach has not yet been presented. Rickenbacher et al. [31] introduced a cost model for L-PBF that considered only some components to calculate the manufacturing costs. However, they neglected some relevant items, such as the fixed cost of the machine due to the maintenance and the heat treatment required to release the thermal stresses. Moreover, they introduced arbitrary factors to model the frequency of material changes, which was evaluated on the basis of a single build rather than a single part. Baumers et al. [32] proposed a general production cost model for electron beam melting (EBM) and direct metal laser sintering (DMLS), which is the EOS GmbH tradename for their L-PBF process machine. Their study estimated the costs according to machine usage. However, they did not consider any design optimizations and neglected the partition of the machine cost of when different components are produced in the same job. A general estimation model of the manufacturing costs of AM processes should consider the so-called well-structured costs, $C_{well\text{-}structured}$, which cover the direct and the indirect costs and can be computed for a single produced part. The direct costs refer to the costs that are directly associated with the production. They are absent if the production is halted. The indirect costs refer to those costs which cannot be avoided when the production is interrupted, such as the salaries of the administrative staff. The direct costs and the indirect costs are functions of the build time of each part.

As far as the indirect costs are concerned, the following costs were calculated:

- Machinery depreciation, which is distributed over the total working hours of the year and computed in proportion to the build time, according to Equation (1):

$$S = \frac{C_{machine}(1+i)^{n}}{n} \cdot \frac{t_{build}}{h_{year}} \tag{1}$$

where $C_{machine}$ is the cost of the machine, n is the number of years, which is usually assumed equal to 5, i is the interest, h_{year} is the annual working hours, and t_{build} is the building time.

- Administrative costs, which are distributed over the total working hours per year and computed considering the build time;
- The costs due to the renting the premises are computed considering the space occupied by the machine and the auxiliary systems.

The direct costs are:

- The design costs per part C_d, which can be referred to as the time required to design and optimize the geometry. Since an optimized geometry must be obtained from a manufacturing design, C_d also refers to the time spent assembling the job (all activities included in checking Figure 1 such as the creation of the STL file, orientation, creation of the support structures, slicing and setting the process parameters). Thus, C_d can be computed as:

$$C_d = \frac{1}{N_{ps}}\left[\left(C_{doper} + \frac{C_{CADsw}}{h_{CADsw}}\right)t_d + \left(C_{doper} + \frac{C_{CAMsw}}{h_{CAMsw}}\right)k_1 t_{job}\right] \tag{2}$$

where C_{doper} is the designer's hourly rate, expressed in €/h, C_{CADsw} and C_{CAMsw} are the cost per user of the annual software license for the CAD model and the job preparation, respectively, h_{CADsw} and h_{CAMsw} are the number of hours of use of the software per year (€/year) for the CAD model and the job preparation, respectively, t_d is the time that is required for the design and t_{job} is the time that is required to prepare the job, and it is weighed by k_1, which is the ratio between the volume of the part (including the support structures and the allowances) V_{O+A+S} and the total job volume V_{job}. k_1 is used to account the building of parts with different geometries in the same job.

k_1 considers that the larger the volume of the part, the more time is required for the building time. N_{ps} is the total number of parts that have to be produced.

- Setup Cost per part, which refers to the preparation of the machine before the job starts. It includes the cost of filling the dispenser, C_{fill}, the cost of preparing and checking the chamber, C_{env} (e.g., argon flow), the cost of resurfacing the build platform, and $C_{build\ plat}$, the cost of removing the supports after the part has been removed. Build platforms are usually re-surfaced by milling operation [33]:

$$C_{setup} = k_2 C_{fill} + k_1 C_{env} + k_3 C_{build\ plat} \tag{3}$$

where $C_{fill} = C_{oper} \cdot t_{fill}$, and t_{fill} is the time required to fill the dispensers. $C_{build\ plat} = (C_{machining} \cdot W(H + ex)) / (0.50 D v_o)$ is the cost of resurfacing the build platform by machining in a single operation. $C_{machining}$ is the hourly cost of the milling machine, W and H are the dimensions of the build platform, ex is the sum of the approach length and the overtravel, D is the diameter of the mill, and v_o is the material removal rate (mm/min). As k_1, k_2, and k_3 take into account the building of parts with different geometries in the same job, k_2 considers the percentage of material utilized for the building of the part against the quantity used to fill the dispensers ($Mat_{filldisp}$). $Mat_{filldisp}$ considers a quantity of powder expressed in kg that corresponds to the quantity of material to fill a building volume corresponding to the maximum height of job multiplied for the dose factor. The dose factor depends on the saturation of the build platform. An additional 20% of material could be also considered. k_3 considers that the larger the projected area of the part on the build platform is, the higher the cost of the part for the milling operations:

$$k_2 = \frac{W_m}{Mat_{filldisp}} \tag{4}$$

$$k_3 = \frac{\text{surface of the build platform occupied by the part}}{\text{total surface of the occupied build platform}} \tag{5}$$

where W_m is the quantity of material used to build the part and $Mat_{filldispis}$ the quantity of material used to fill the dispenser.

When Argon is used, C_{env} can be computed as follows:

$$C_{env} = C_{Ar} V_{Ar} + C_{oper} t_{operAr} \tag{6}$$

where C_{Ar} is the price of the Argon per m^3, and V_{Ar} is the total volume of Argon used to fill the build chamber and achieve the right pressure before the process starts. t_{operAr} is the time required for the operator to start and control the procedure.

- Production cost, $C_{production}$, refers to the direct cost of building the part. This cost includes the energy consumption of the machine and of the other systems, including gas consumption. In addition, this cost includes the maintenance of the machine and the other systems during which a downtime period is required. These costs are computed as indirect costs and are a function of the time that the machine is used to build the part. However, these costs are not taken into account when the machine is not utilized, unlike the indirect costs, which still have to be considered, even when the production is halted:

$$C_{production} = C_{gas}\left(t_{exp} + k_4 t_{cooling}\right) + C_{AM} t_{build} \tag{7}$$

where C_{gas} is the hourly rate cost of the gas that is used during the building and cooling of the part, C_{AM} is the direct hourly rate cost of the AM system and is the sum of the costs of the energy consumption per hour, of the maintenance of the machine and the other C_{mm} systems.

C_{mm}, which is distributed over the total hours between two subsequent maintenance operations of the t_{b2m} systems, is calculated as follows:

$$C_{mm} = \sum \frac{\left(C_{oper} t_{mm} + C_{rc}\right)}{t_{b2m}} \quad (8)$$

where C_{oper} is the operator's hourly rate cost, expressed in €/h, t_{mm} is the time that is required for the maintenance operations and C_{rc} is the cost of the replaced components.

- Material cost, C_{mat}, is obtained from the total material quantity W_m and is calculated according to Equation (9). The cost of powder, C_{powder}, refers to the cost per kilogram of powder:

$$C_{mat} = W_m C_{powder} \quad (9)$$

- The manufacturing cost is related to the build time, and it includes the idle time and the exposure time, t_{exp}, per part, as presented in Equation (10):

$$t_{build} = k_1\left(t_{heating\ plat} + t_{aux} + t_{fillAr}\right) + t_{exp} + k_4 t_{cooling} \quad (10)$$

where $t_{heating\ plat}$ is the time required to heat the build platform, t_{aux} is the extra time required before starting the process (the cleaning and levelling processes of the build platform, compacting and leveling the powder, cleaning the lens and lens cover), t_{fillAr} is the time required to fill the build chamber with argon, and $t_{cooling}$ is the time required to cool the part. k_4 is a coefficient that is introduced to consider the building of parts with different geometries in the same job and is defined as follows:

$$k_4 = \left(1 - \frac{\text{the total surface of the part}}{\sum \text{the surface of the parts}}\right) \quad (11)$$

k_4 considers that the larger the surface of the part is, the more rapid the cooling.

- The post-processing cost, $C_{post\ proc}$, only includes the operations that are mandatory to consider the AM process complete. For these reasons, $C_{post\ proc}$ contains the cost of removing the support structures, $C_{rem\ supp}$, the cost of the post treatment, $C_{post\ treat}$, the cost of the heat treatment to release the residual stresses for the L-PBF process, and the cost of polishing the part, $C_{polishing}$, by shot peening:

$$C_{post\ proc} = k_3 C_{rem\ supp} + k_1 C_{post\ treat} + C_{polishing} \quad (12)$$

$$C_{rem\ supp} = (C_{EDM} v_{oEDM} + C_{saw} v_{oGr}) l_{build\ platform} + C_{oper} t_{rem} \quad (13)$$

The support can be removed by means of a wire electro discharge machining (EDM) process, manually or by sawing. C_{EDM} and C_{saw} are the hourly costs of the EDM machine and of the saw, respectively. v_{oEDM} and v_{oGr} are the material removal rates for EDM and sawing processes, respectively. $l_{build\ platform}$ is a length of the build platform and t_{rem} is the time required to remove the support structures manually. $C_{post\ treat}$ and $C_{polishing}$ are evaluated as the hourly costs for the machine and the time needed to complete the operation. k_3 considers that more supports are necessary for larger surfaces. k_1 considers that the larger the parts are, the longer the time needed in the oven.

- Finishing costs, which refer to the additional operations necessary to finish the part and achieve the required dimensional, geometrical, and surface accuracy. This information should be defined at the design stage.

4. Case Study

The study has dealt with the system (Figure 2) that supports and moves a flying measure probe in a working volume. Each test machine has eight flying measurement probes: four to acquire signals from the top of the board and four to acquire signals from the bottom of the board. The high precision flying probe is supported by a bracket which also contains the vision system and the lighting system that are used to acquire images during the measurements, the mechanisms that are used to move the flying measure probe and the data collection systems. The flying probe and the system used to collect data and to move the probe are joined to the bracket by means of six screws, while the vision system and its data collection system are joined to the bracket with four bolts. The whole system (bracket, flying probe and vision system) is joined to the X-rail by four bolts and aligned precisely with two dowel pins. The linear motor guide is assembled in the upper part of the X-rail. The X-rail also contains the motor stator magnets which are enclosed in the rail by two side covers. The X-rail is joined to the bottom part of the Y-rail and is equipped with another linear motor. The electric motors move the whole system linearly along the X and Y axes. The probe can also be moved along the Z*axis, which is rotated by a certain angle with respect to the normal of the XY-plane. The overall dimensions of the two rails (orange in Figure 1) are 220 × 690 × 101 mm^3 and the total weight is around 5000 g. The maximum envelope for the bracket is 150 × 60 × 60 mm^3. The bracket weighs 203 g, while the whole system (probe, vision and lighting systems) weighs about 1400 g.

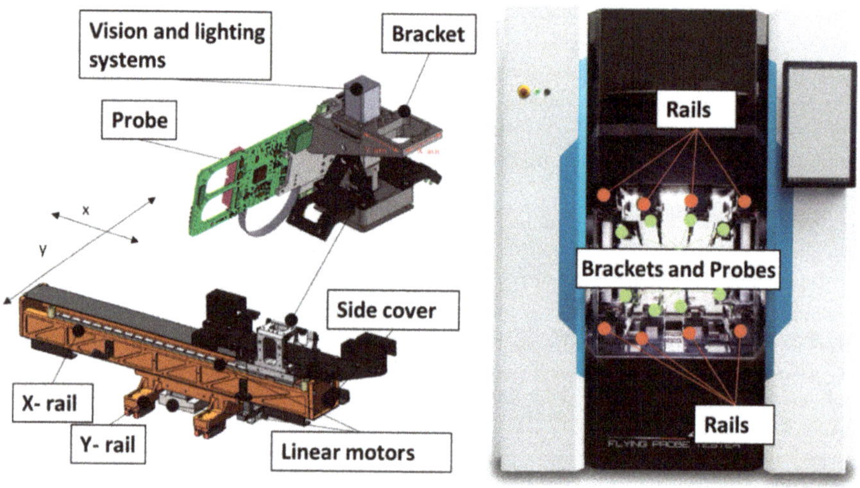

Figure 2. Case study. The systems that are subjected to redesigning are the X and Y rails and the bracket.

From the structural point of view, the most critical components for the accuracy of the machine and its dynamics are the two rails and the bracket. Because of the low number of parts produced per year (around 400 parts), a three-axis milling machine is used today to shape the components from raw aluminum 7075 alloy ingots.

The part of the system that considers the two rails includes 16 elements and requires 12 operations to join the two rails. The weight of these elements lies on the linear motor and affects the acceleration and deceleration ramps when the flying probe is moving in the working volume. Owing to the design criterion which lies to the high stiffness, the system is subjected to low stresses. The redesigning of the rails is aimed at streamlining the assembly in order to reduce the geometric errors that are accumulated and propagated, step by step, during the assembly process and which may affect the machine accuracy [34]. Redesigning is also aimed at reducing the total weight while maintaining stiffness. The design limitations, due to the larger rail dimensions (690 mm) than for most industrial metal component AM systems, need to also be considered at the design stage together with subsequent machining operations.

The current design of the bracket exhibits a maximum displacement of over 30 μm, which exceeds the design requirements of 20 μm. The generated bending excess affects the position of the probe and the precision of the measurements. This error is currently compensated for by adopting suitable algorithms to adjust the probe positioning. From an industrial point of view, this choice is the best compromise between the time and costs necessary to design and manufacture (machine and component set, tools, etc.) the bracket. The bracket redesign is aimed at increasing the stiffness to that of the requirement (20 μm) while maintaining its original weight to avoid an increased load on the electric motors.

Both components are designed to be produced using gas atomized AlSi10Mg powder, as it has similar properties to the original material. The material properties are reported in Table 1. Exact details about the machine, the geometries, the load conditions and the working cycle considered in this case study cannot be disclosed for confidentiality reasons. However, this should not rule out the understanding of the main findings of the case study.

Table 1. Material properties: comparison of Aluminum 7075-T6 for the three-axis milling machine and AlSi10Mg for L-PBF systems.

Property	Aluminum 7075-T6	AlSi10Mg
Density [kg/m^3]	2810	2670
Young's Modulus [MPa]	71,700	75,000
Yield Strength [MPa]	503	350

4.1. Design of the Rails

A simplified model of the original system was initially implemented in Inspire™ 2018 SolidThinking®, version 2018, build 9508, to analyze its mechanical behavior. The probe, and the vision and the lighting systems were modelled as concentrated masses. The magnetic forces from the linear motor were simulated as vertical loads. A finite element (FE) method was used to solve the model. As expected, the system showed an almost uniform stress distribution and high equivalent stress of around 20 MPa, which is far below the limit of the material (Table 1). As already mentioned, the low stresses are due to the high stiffness design criterion. Thanks to the low stress (see supplementary file), the part can be redesigned freely considering only the peculiarity of the L-PBF process. The components of the system were therefore redesigned to achieve an efficient new material distribution and weight reduction while only coupling areas were constrained.

As far as the X-rail is concerned, the design starts with a redesigning of the vertical walls (Figure 3a,b). Each vertical wall was replaced by two thin honeycomb-structure walls. These structures were demonstrated to be self-supported, and allow better stress distribution and light. The walls were also placed out of sync in order to distribute the stresses more evenly during the production of the components and during operation. A new material distribution was considered for the bottom wall using the Voronoi algorithm (Figure 3c). The cover was then welded to the rail (Figure 3d) to improve the stiffness of the rail and it was shaped like a cross. A similar approach was followed to redesign the Y-Rail (Figure 3e).

After the redesign of every single part, the parts were assembled and unnecessary features were removed. The whole system was therefore redesigned to be produced as a single monolithic part as follows. The two rails were then welded together. However, the component had to be split into three parts to fit most commercial and industrial L-PBF systems. Owing to the dimensions of the new parts, with the aim to produced them in a single job, an EOS M400 machine (400 × 400 × 400 mm^3) has been considered for the production. Figure 4a shows the results of the optimization phase (Section 2 and Figure 1) in which the support structures and the surface finishing are minimized. In this way, only the top surfaces of the X-rail, and the bottom surfaces of the X and Y rails have to be finished. The bottom surfaces are finished directly by EDM when the parts are removed from the building platform. The top surfaces of the X-rail are machined by milling after the assembly of the parts to ensure a good geometrical tolerance so that they can be joined to the linear motor guides. Figure 4 shows a simulation of the fastening systems during a milling step.

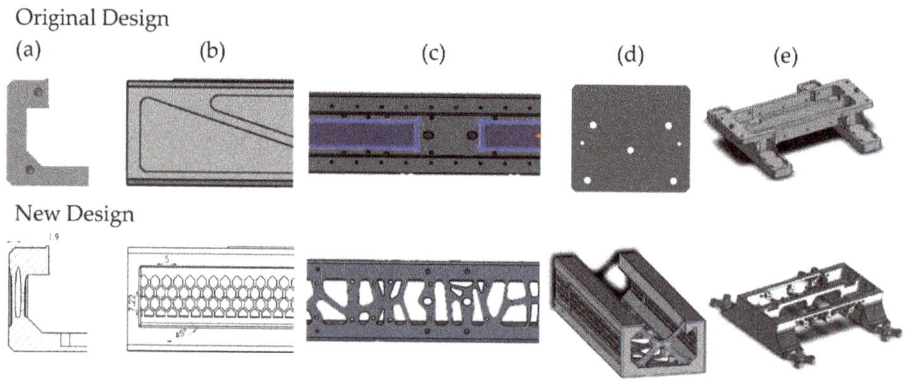

Figure 3. Details of the redesigned features for (**a–d**) the X-rail and (**e**) the Y-rail.

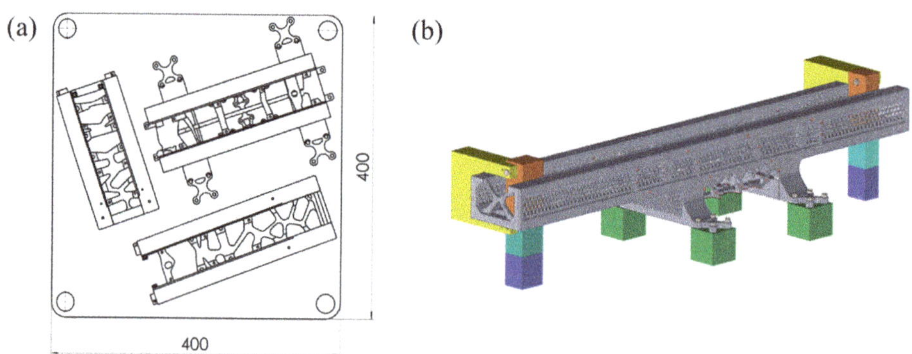

Figure 4. (**a**) manufacturing orientation for L-PBF production by EOS M400 and (**b**) machining using a three-axis machine.

According to the support optimization indications (Section 2 and Figure 1), all surfaces were modified to avoid the need of support structures (e.g., Figure 5a) and where the geometrical features were not accessible for manual or mechanical support removal (e.g., Figure 5c, an internal section of the y-rail). A grid (Figure 5d) that also works as a support during the L-PBF process was designed to increase the stiffness of the systems. A shaft-hub interference fit was designed (Figure 5e) to assemble the three parts (details provided in the supplementary file). The connection only included features that are self-supported. The connection between the two parts was also ensured by two bolts (Figure 5b). The component parts were all numbered to make the assembly easier (e.g., Figure 5a red square). After the geometry modification, Figure 6 shows the result of the support optimization according to Section 2 and Figure 1.

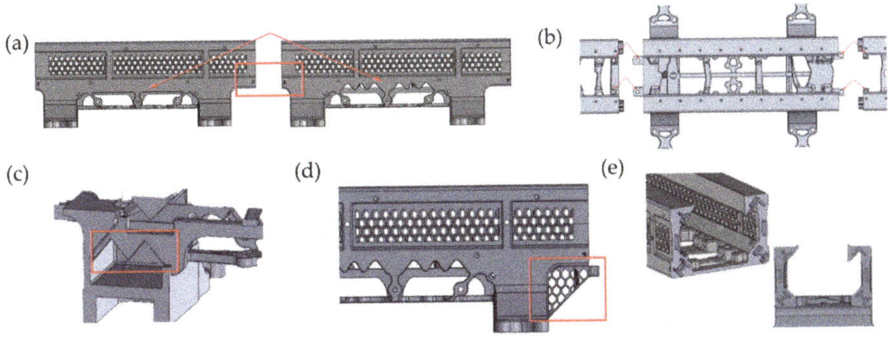

Figure 5. Modifications of the part features according to the support optimization procedure presented in Section 2. (**a**) modification of the bottom part of the rail to reduce the number of support structures; (**b**) removing material and modification of the surfaces to avoid inaccessible area for the support removal (a section is showed in (**c**)) and adding a link to improve the assembly strength; (**d**) features that support the overhang both during the construction of the part and the working conditions; (**e**) particulars of the shaft–hub interference fit (details are provided in the supplementary file).

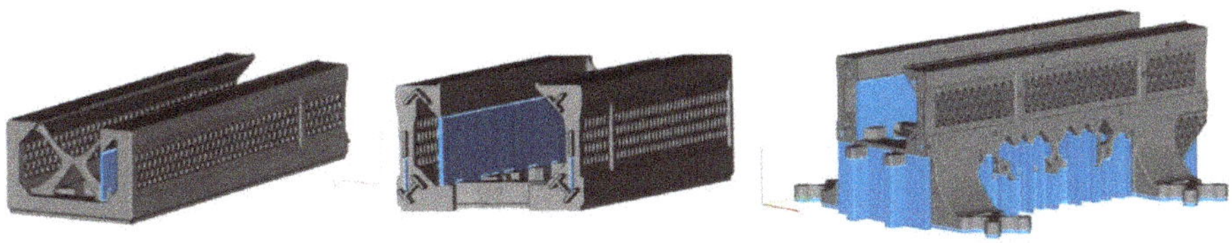

Figure 6. The layout of the parts with the support structure (in blue). All supports can be removed manually, except the ones on the bottom surfaces which are removed during the detachment of the parts from the building platform by EDM.

Figure 7 shows the differences between the original and the new design. The displacements showed the same distribution. The new material distribution that led a reduction in weight of about 32% was achieved as a result of the implementation of the new design, without any significant changes in terms of stiffness. A technical prototype for the experimental tests was produced according to the designed working cycle. The total production time, including the heat treatment, the support removal, the EDM, and the finishing operations, was about three days. The final component is shown in Figure 8. The total material swarf was about 200 g. The material cost (powder cost equal to 65€/kg) of producing the component was around €197, which is comparable with the material cost of the original component (€196 [35]). However, considering that the initial components were machined from an ingot [35], the material saving, with respect to the original component, was around 99%. The assembly operation flow was simplified because all the operations to join the rails and the covers were removed.

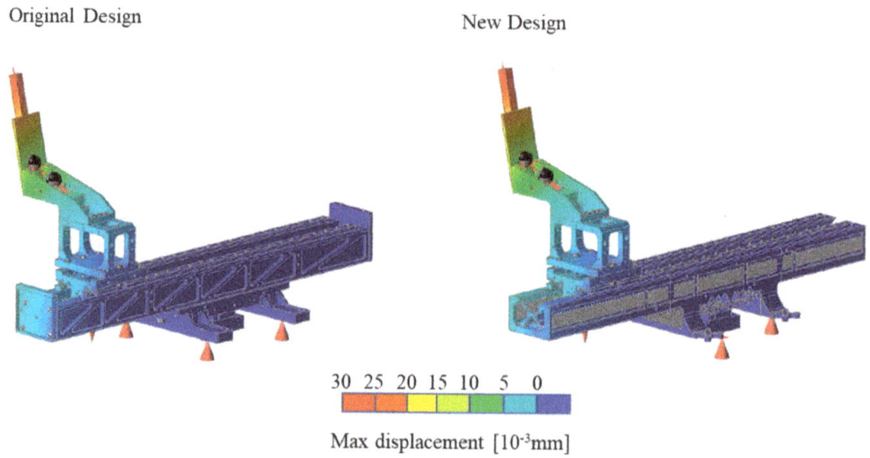

Figure 7. Comparison of the maximum displacements between the original and the new design. The maximum displacement is registered as the extremity of the probe. The slight difference in the displacements values between the two designs demonstrates that the stiffness of the system has been preserved while the material efficiency has been improved.

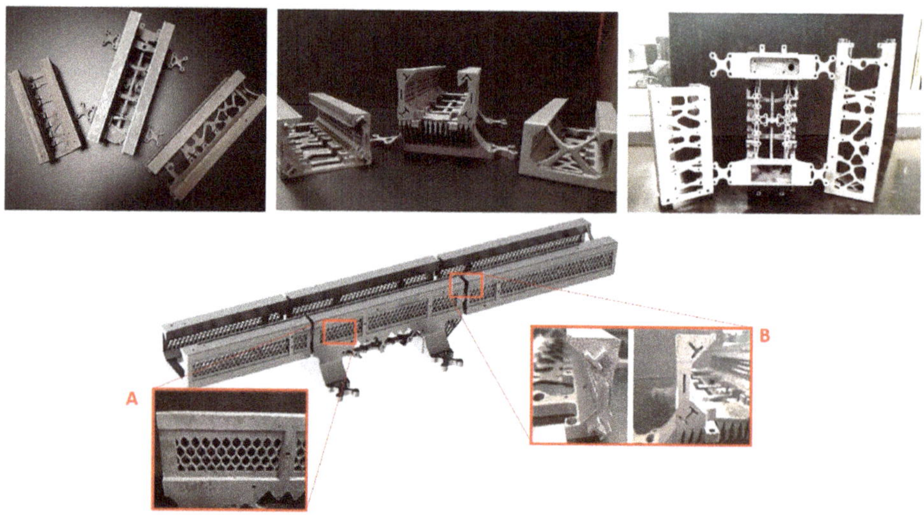

Figure 8. Technical prototype with details of the walls of the X-rail (**A**) and the shaft-hub interference fit (**B**).

4.2. Bracket Design

As already mentioned, the redesigning process had the aim of designing a bracket that would satisfy the original requirement, in terms of displacement, while maintaining its original weight. In order to maintain the original rail assembly operational flow, the coupling areas were constrained with other systems to avoid modifications as a result of the redesign. Some other surfaces were also included as constrained areas so that the same fastening tools used for the original geometry could be used during the finishing operations.

Differently from the previous case, an increase of the stiffness is required here. Practically, this means a search of a new material distribution which considers a larger design domain of the original component. The design domain needs to be large enough to include the anticipated optimal material distribution, but also had to be as small as possible to avoid unfeasible structures and increases in the computational costs due to an abundance of unnecessary elements [36]. The potential design domain has been considered as the design space that does not compromise the fastening operations with the other systems. The design domain was therefore increased and optimized iteratively according to the following procedure:

1. Defining the initial design by increasing the updated design space without exceeding the maximum envelope:
   - Identifying the areas with high deformation energy from the computer-aided engineering (CAE) analysis of the original design;
   - Adding material to the areas that showed high deformation energy, after controlling that the total volume did not increase excessively. It noticed that, for small components, the total volume should not have exceeded twice the updated volume in order to avoid an excessive increase in the computational time. This rule was therefore followed in the subsequent steps.
2. Meshing the domain with a tetrahedral mesh which could be generated automatically, in order to decrease the computational time. In this step, the density of the mesh could be low.
3. Running a TO with free constraints and using the solid isotropic material with penalization (SIMP) method [37], which is implemented in the most frequently used commercial software programs, by constraining the weight to be equal to that of the original design (or a volume reduction of 50%) and maximizing the stiffness.
4. Updating the optimized geometry (i.e., creating a geometry in which the areas with low deformation energy had been removed).
5. Verifying that the maximum stress did not exceed the material limit.
6. Establishing the maximum displacement of each axis; if the values did not exceed the prescribed condition, then the geometry obtained from step 4 could be considered optimal; otherwise, it was necessary to go to the next step.
7. Modifying the geometry obtained from step 4 by adding material to the areas with high deformation energy, while controlling that the total volume did not exceed twice the original volume. This model represents the new design space.
8. Go back to step 2.

The as-defined design space results in being the smallest that would allow a solution to be found for the global displacement. In short, the proposed design space optimization procedure solves a displacement control problem by means of a series of structural TO problems in which the minimum compliance and volume constraint are considered. The maximum displacement is considered as the ideal design optimization criterion for the design space. Therefore, the design space is redefined several times, and the optimized geometry is obtained after iterations. These iterations work as a design space optimization as they allow the domain to be expanded, where necessary, regardless of the shape of the initial design domain. In other words, the displacement control problem is solved by means of iterative lightening procedures, in which the optimized goal is to find the structures with the maximum stiffness. That obtained design space design space was used in a free constraint TO using an improved mesh quality to provide a more precise solution [7] and allow a better representation of the structure.

The design optimization of both initial design domain and final geometry have been obtained by the TO method. The TO and validation of the optimized geometry were implemented and solved using Abaqus Standard and a TOSCA algorithm. Thanks to the design freedom, which is guaranteed by the design for the L-PBF process, TO was run with free constraints [38] and using the SIMP method [37].

A tetrahedral mesh was used for the design space optimization and for the final TO in order to decrease the computational time. Hexahedral elements were used for the analysis and validation of the geometry because the accuracy of the results was considered of utmost importance.

4.2.1. Redesigning and Prototyping

A preliminary static analysis of the original design of the bracket was performed to validate the FE model for the stiffness optimization. The analysis involved the implementation of an FE model of the original bracket, in which the mechanical stress and displacement behaviors were included. The probe and the vision system weights and the probing force during the measurement are the main causes of the bending of the bracket. The bending of the bracket was thus simulated for two different load cases. Load case 1 represents the situation in which the bracket is subjected to its own weight and to that of the other systems (vision and lighting systems and probe), while load case 2 represents the phase in which the probe tests a MEMS relay, and a probing force is therefore added. The maximum equivalent stress on the component was around 15 MPa which was still below the limit of the material so as not to decrease the stiffness of the component. However, the maximum equivalent displacement was around 35 μm, in agreement with the specifications provided by the manufacturer. According to the procedure mentioned above, the design space was subjected to several iterations. An expansion of the original design is possible if the geometry modifications do not lead to a change in the finishing and assembly operations. The starting point was the original design for which the stresses and the displacements were known (Figure 9).

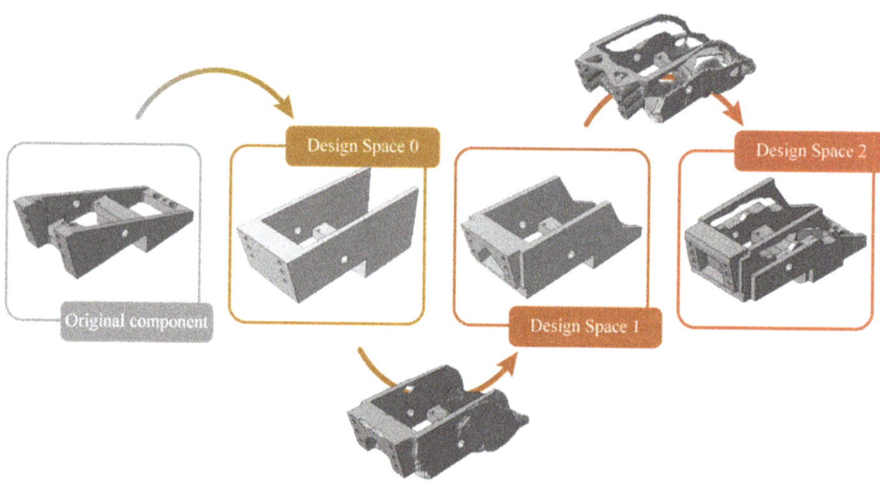

Figure 9. Iterations of the design space optimization. At each iteration, the material has been removed from the area at low energy deformation and has been added in the ones at high energy deformation. The iterations ended when the maximum displacement of the bracket under the load was less than the prescribed requirement (20 μm).

Design space 0 was obtained by adding material to the areas with high deformation energy (Figure 9) while controlling that the total volume did not exceed twice the original volume. TO was then run by constraining the weight to be equal to the original design and maximizing the stiffness. The stresses and displacements were then analyzed. Design spaces 1 and 2 (Figure 9) were obtained from the previous design space in order to consider the optimized geometry by adding material to the areas with high deformation energy while controlling that the total volume did not exceed twice the original volume. Finally, when TO was run using design space 2, the displacement values did not exceed the prescribed condition, that is, 20 μm and the space design was therefore considered optimal. Each design space was using Abaqus CAE and re-meshed considering the same element size. Design space 2 was then used to perform a more detailed optimization. The element was decreased in size to improve the approximation of the geometry.

The optimized geometry was redesigned, according to Section 2. Figure 10 shows a comparison of the original and the redesigned geometry displacements, which are reduced to the prescribed constraint in each direction. The maximum displacement was along axis 2 and was equal to 18 μm. Furthermore, the final design weighed 184 g, which is about 10% less than the original bracket. This reduction in weight, in AM material, cannot be ascribed to the slightly lower density (Table 1) because the difference between the material density values is about 5%.

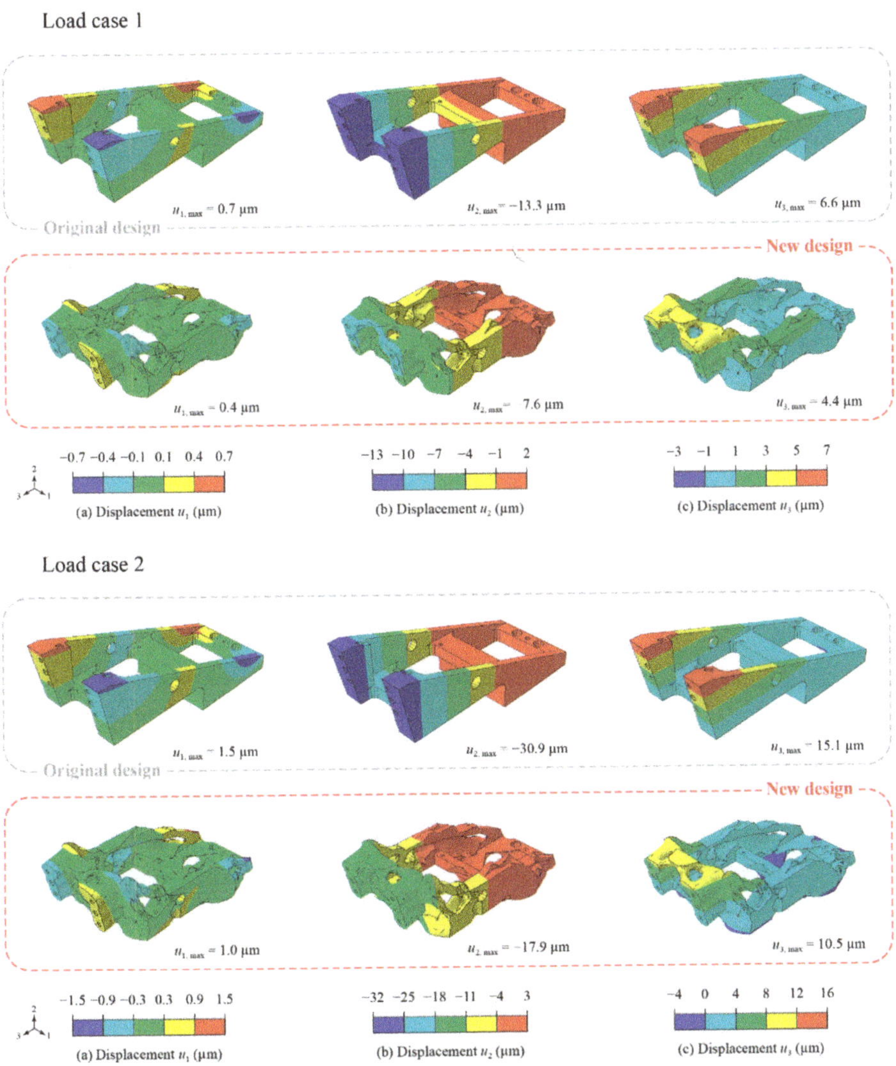

Figure 10. Comparison of the original and new designs for load case 1 (the bracket is subjected to its own weight and to that of vision and lighting systems and a probe) and load case 2 (the probe tests a MEMS). The lower displacements in the case of the new design demonstrated an overall increasing of the stiffness of the bracket.

The feasibility of the bracket in terms of part and process (choice of supports and orientation) designs has been verified by producing a single part. With this scope, the production has been performed by an EOSINT M270 Dual Mode machine that has a small production volume (250 × 250 × 215 mm^3) which is suitable for the preliminary prototyping tests. An Ytterbium fiber laser system is used to melt powders with a continuous power of up to 200 W, a spot of 100 μm, and scanning rate up to 7000 mm/s in an argon atmosphere. The used process parameters are shown in Table 2. The produced component was heat-treated in a furnace (2 h at 300 °C) to prevent inaccuracies due to stresses induced thermally during removal of the parts. Figure 11 shows the as-built technical prototype in AlSi10Mg manufactured by the EOSINT M270 machine.

Table 2. Process parameter values employed for the bracket production.

Process Parameters	Skin	Core	Contour
Scan speed [mm/s]	900	800	900
Laser power [W]	120	195	80
Hatching distance [mm]	0.10	0.17	-
Layer thickness [μm]	30	30	-
Laser spot size [mm]	0.10	0.10	0.10

Figure 11. The technical prototype produced by EOSINT M270.

4.2.2. Cost of the Bracket

A production series of the bracket with the new design has been considered to compare the feasibility of the product and the actual cost with those of the original design. A production quantity of 400 parts per year was considered [35].

The time to produce the original design, using a milling machine, is about 1.40 h, and the total cost is around €116 [35].

A SLM500 machine, made by SLM Solution, was considered for the simulation because, among the industrialized L-PBF systems, its building volume fits in a single job a higher number of brackets. The build volume is 500 × 280 × 365mm^3, where 500 × 280 mm^2 are the dimensions of the build platform (WxH). The price of the machine is around €1,200,000. The powder necessary to fill the tank ($Mat_{filldisp}$) is about 50 kg in which a maximum height of the job and a dose factor equal to 27.3 mm and 4, respectively, have been considered. The manufacturing costs were simulated by considering the build platform to have been filled according to the procedure reported in Section 2. Nine parts can be produced for each job. An aliquot of 0.240 kg of material is necessary to build the part and supports (W_m). An additional 5% is considered as powder lost during the cleaning of the part. After the production, the loose powder is removed from the parts and the as-built parts (attached to the building platform) are heat-treated to release the residual stresses. The supports are then removed manually. The pinholes and the holes for the bolts that join the probe to the bracket are the only surfaces that must be finished. It was assumed that all the manufacturing operations were performed in the same workshop.

The above-mentioned data are summarized in Table 3. The coefficient k_1 is calculated by the ratio between the V_{O+A+S} and the total job volume V_{job}, k_2, and k_3 are calculated according to Equations (4) and (5), while k_4 is equal to zero because no cooling time is required after the production. The times included in Table 3 that are necessary to complete any manual operation or task was measured in real production. With this scope, a single part was produced using an SLM500 (Figure 12).

Table 3. General information about the cost of the model.

Material quantity for the build tank [kg]	$Mat_{filldisp}$	50
The dimension of the build platform [mm^2]	WxH	500 × 280
Volume (part+supports) [mm^3]		90,000
Material quantity (part+ supports) [kg]		0.240
% of waste		5
Total material quantity [kg]	W_m	0.240012
Number of parts per job		9
k_1		0.11
k_2		0.005
k_3		0.11
k_4		-
Operation time		
Heating platform [h]	$t_{heating\ plat}$	0.33
Levelling platform [h]	t_{aux}	0.33
The time necessary to fill the chamber with Argon [h]	t_{fillAr}	1.67
The time necessary to complete the job [h]	t_{exp}	13
Cooling time [h]	$t_{hooling}$	-
Total time per part [h]	t_{build}	1.7

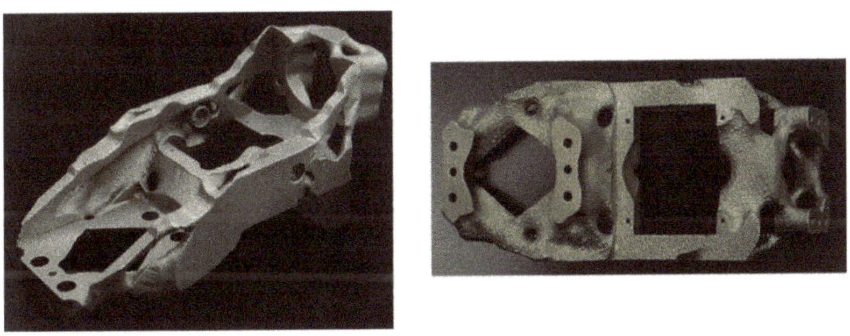

Figure 12. Part manufactured by SLM500. The idle times (Table 3) have been measured during the production.

Table 4 shows the calculation of each costs. The cost for a single part is obtained by dividing the cost for the number of brackets that are fabricated in a single job. The machining times were numerically calculated performing a 3-axis milling using computer-aided manufacturing (CAM) software, Visi 19. The same clamps used for the original designed were used. The costs of designing the part and the job were neglected because a large production was considered.

Table 4. Calculation of the total manufacturing costs according to the model presented in Section 4.1. * The volume of Argon necessary to fill the chamber, V_{Ar}, is the product of t_{fillAr} and Ar_p.

Machinery Depreciation and Indirect Costs		
Cost of the machine [€]	$C_{machine}$	1,200,000
Years		5
I		0.05
Depreciation cost [€/h]	S	76.58
Indirect costs [€/h]		11.5
Indirect costs per part [€/part]		149.74
Setup Costs		
Labor cost [€/h]	C_{oper}	23
The time necessary to fill the dispenser [h]	t_{fill}	0.7
Cost of filling the dispenser [€/part]		0.010
Time of the operator to start the process [h]	t_{operAr}	0.08
Cost of the Argon [€/m^3]	C_{Ar}	3.7
Argon consumption for purging [l/min]	Ar_p	70
Total Argon volume to fill the chamber * [m^3]	V_{Ar}	7.03
C_{env} [€/part]		3.09
Machining cost [€/h]	$C_{machining}$	35
Labor cost [€/h]	C_{oper}	23
Mill diameter [mm]	D	100
Cost of resurfacing the building platform [€/part]	$C_{build\ plat}$	2.05
Setup cost per part [€/part]	C_{setup}	5.15
Material Cost		
Powder cost [€/kg]	C_{powder}	65
Material cost per part [€/part]	C_{mat}	19.06
Production Costs		
Cost of the Argon [€/m^3]	C_{Ar}	3.7
Argon consumption during the building [l/min]	Ar_b	5
Total Argon volume for building [m^3]	V_{Ar}	0.433
AM cost [€/h]	C_{AM}	10
Production cost per part [€/part]	$C_{production}$	18.60
Post-processing costs		
Labor cost [€/h]	C_{oper}	23
EDM cost [€/h]	C_{EDM}	30
The time necessary to remove the supports manually [h]	t_{rem}	1
Heat treatment cost [€/h]		35
Heat treatment time [h]		8
Polishing cost [€/h]		5
The time necessary to remove the powder [h/part]		0.16
Post processing cost per part [€/part]	$C_{post\ proc}$	43.69
Manufacturing cost [€/part]		247.59
Finishing operation cost [€/part]		52.7
Total cost [€/part]		**300.37**

The manufacturing cost for the new design (€300.37) results in being more than twice those of the original bracket. Figure 13 compares the distribution of each cost item over the total manufacturing costs for the new design and the original one. It may be observed that the largest contributions to the AM processes are those of the depreciation of machinery and the indirect costs. This is because these costs are computed on the basis of the build time. In fact, the production time of L-PBF machines is longer than that of the traditional manufacturing process. The depreciation of machinery and the production costs for a 3-axis CNC (computer numerical control) process are distributed equally over the total costs. In fact, the total hourly production cost for a traditional machine considers the maintenance of the machine, the tools, the equipment, the lubrication, the cooling system and energy consumption, the operator, movement of the pallet, and so on. As a result, the total hourly production cost for a traditional process is higher than that of an AM process. A large part of the cost of the new design is due to the post-processing and finishing operations. As the currently used L-PBF processes can only produce near-net shaped parts, the total production cost of obtaining a functional part, by means of L-PBF, should be considered as the sum of the production cost, the post-processing and the finishing cost. Accordingly, it is evident that the distribution of the costs of traditional and AM processes is similar. Although it is generally believed that the price of powders influences the final cost of the part to a great extent, the here presented analysis instead shows that the main influence on L-PBF is the production rate, in other words, the time necessary to produce a part and to finish it. The cost of the material to produce the original design, by means of milling, is comparable with the cost of the material necessary to produce the newly designed part produced by L-PBF. However, the cost of the material for 3-axis CNC has a significant effect on the total cost because a large quantity of material is wasted during machining. In fact, a 2.2 kg ingot of material is needed to produce a bracket that weighs 203 g.

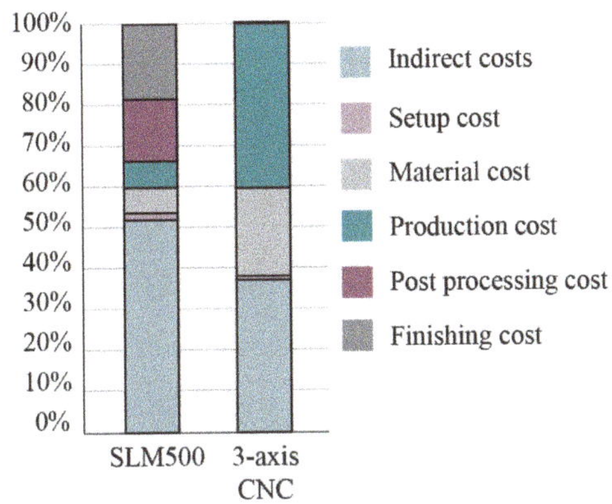

Figure 13. Distribution of the manufacturing costs.

Besides the increased cost, the benefit that was achieved by adopting the new geometry, and thus the AM process adopted to produce the part, should be considered from an overall point of view. The new design for L-PBF has shown the possibility of achieving the required stiffness, which in turn improves the actual machine accuracy and dynamics because the load on the electric motor is decreased. Therefore, the higher cost, with respect to the 3-axis CNC, could be justified by the increase in the performance of the component and in turn of the whole machine. The same conclusion cannot be reached when the new design and the 3-axis CNC process are adopted because the complexity of the geometry would lead to an exponential increase in manufacturing costs.

5. Conclusions

The here presented work is aimed at exploiting the production advantages of L-PBF systems for high-precision applications, in which the design problems pertain to the assembly operations, weight reductions and increasing the stiffness. In this context, the rails and bracket that support a high-precision flying probe for a testing machine have been redesigned. The redesign is based on a general methodology that considers all the steps of an L-PBF process. The optimization of the design, the support and the part orientation have been driven by material efficiency. The components that have been redesigned to be produced by L-PBF have demonstrated great advantages, in terms of mechanical performance. The weight of the rails has been reduced by 32%, while the number of components has been reduced from 16 to 3. The optimized bracket results in being more than 50% stiffer than the original one and about 10% lighter. In both cases, CAE simulations have shown that great improvements can be achieved as a result of new material distribution. The results have thus provided further validation of the fact that an integrated approach between design and AM processes is an extremely powerful tool to generate geometries with high performances and characteristics, even in the case of precision machines, coordinate measuring machines, and industrial robots, in which the mechanical stiffness of the parts, without any extra weight, is one of the most important design criteria. The cost analysis has also confirmed the suitability of adopting a new manufacturing process to produce components for this industrial sector. The higher cost of the new design of the bracket, with respect to the original one, appears negligible compared to the improvements that can be achieved in terms of component performance and machine accuracy. Additionally, the increased cost appears to be a small part if compared with the market price of these kinds of machines (around €700,000). Compared with the traditional process, DfAM leads to significant savings in material. As a result, the use of AM leads to positive environmental impacts, in terms of material utilization, the supply chain, and life-cycle performances [39].

Overall, AM processes shorten the distance between concept and reality. However, the empirical findings of this study suggest important issues that should be considered in future research:

- The support structures and finishing operations still represent a challenge for the full exploitation of the AM process and for the creation of optimized geometries;
- The available commercial tools are still not able to fully support the designer to find optimal design solutions for AM and to handle geometric complexities;
- At present, the application of an iterative approach to designing seems to be the only way to:
    - reach an optimal solution that takes advantage (for specific purposes) of the level of complexity of the design made available as a result of the adoption of AM processes;
    - achieve the production of truly optimized parts.

Supplementary Materials: The following are available online at http://www.mdpi.com/2073-4352/10/3/161/s1, Figure S1: Comparison of the Von Mises stress between the original and the new design. As can be observed, the stress on both parts is extremely low. The maximum values around (24 MPa) is registered in the contact between the bolts the rails (Figure S2). The low stresses and great safety coefficient respect to the yield stress of the material (503 MPa for the Al7075 of the original design and 350 MPs for the AlSi10Mg of the new design) are due to the high stiffness requirements which were applied the design stage (which means low deformation and displacements during the working conditions of the system), Figure S2: Localisation of the only areas in which the maximum stress has been registered, Figure S3: Design details of the shaft-hub interference. The features have been designed to guarantee an adequate stiffness of the connection. This is ensured not only by the feature 'dimensions but also by the roughness of tilted surfaces that is slightly higher of the horizontal and vertical ones and helps to improve the contact between the surface.

Author Contributions: Conceptualization, M.G. and L.I.; methodology, M.G. and F.C.; software, M.G. and M.V.; validation, M.G., F.C. and M.V.; formal analysis, M.G.; investigation, M.G.; resources, L.I.; data curation, M.G.; writing—original draft preparation, M.G.; writing—review and editing, M.G. and F.C.; visualization, M.G. and F.C.; supervision, L.I. All authors have read and agreed to the published version of the manuscript.

Funding: This research received no external funding.

Conflicts of Interest: The authors declare no conflict of interest.

References

1. Gibson, I.; Rosen, D.W.; Stucker, B. *Additive Manufacturing Technologies*; Springer: New York, NY, USA, 2015; Volume 17.
2. Salmi, A.; Calignano, F.; Galati, M.; Atzeni, E. An integrated design methodology for components produced by laser powder bed fusion (L-PBF) process. *Virtual Phys. Prototyp.* **2018**, *13*, 191–202. [CrossRef]
3. Saadlaoui, Y.; Milan, J.L.; Rossi, J.M.; Chabrand, P. Topology optimization and additive manufacturing: Comparison of conception methods using industrial codes. *J. Manuf. Syst.* **2017**, *43*, 178–186. [CrossRef]
4. Biamino, S.; Penna, A.; Ackelid, U.; Sabbadini, S.; Tassa, O.; Fino, P.; Pavese, M.; Gennaro, P.; Badini, C. Electron beam melting of Ti–48Al–2Cr–2Nb alloy: Microstructure and mechanical properties investigation. *Intermetallics* **2011**, *19*, 776–781. [CrossRef]

5. Galati, M.; Iuliano, L. A literature review of powder-based electron beam melting focusing on numerical simulations. *Addit. Manuf.* **2018**, *19*, 1–20. [CrossRef]
6. Brackett, D.; Ashcroft, I.; Hague, R. Topology optimization for additive manufacturing. In Proceedings of the 22nd Annual International Solid Freeform Fabrication Symposium—An Additive Manufacturing Conference, SFF 2011; University of Texas at Austin (freeform), Austin, TX, USA, 17 August 2011; pp. 348–362.
7. Aremu, A.; Ashcroft, I.; Hague, R.; Wildman, R.; Tuck, C. Suitability of SIMP and BESO topology optimization algorithms for additive manufacture. In Proceedings of the 21st Annual International Solid Freeform Fabrication Symposium (SFF)–An Additive Manufacturing Conference, Austin, TX, USA, 9–11 August 2010; pp. 679–692.
8. Rozvany, G.I.N. A critical review of established methods of structural topology optimization. *Struct. Multidiscip. Optim.* **2009**, *37*, 217–237. [CrossRef]
9. Li, B.; Ge, L.; Hong, J. An intelligent computational approach for design optimization of stiffened engineering structures. *Int. J. Precis. Eng. Manuf.* **2017**, *18*, 1005–1012. [CrossRef]
10. Doubrovski, Z.; Verlinden, J.C.; Geraedts, J.M.P. Optimal design for additive manufacturing: Opportunities and challenges. In Proceedings of the ASME 2011 International Design Engineering Technical Conferences and Computers and Information in Engineering Conference, Washington, DC, USA, 28–31 August 2011; American Society of Mechanical Engineers: New York, NY, USA, 2011; pp. 635–646.
11. Seepersad, C.C. Challenges and opportunities in design for additive manufacturing. *3D Print. Addit. Manuf.* **2014**, *1*, 10–13. [CrossRef]
12. Großmann, A.; Weis, P.; Clemen, C.; Mittelstedt, C. Optimization and re-design of a metallic riveting tool for additive manufacturing—A case study. *Addit. Manuf.* **2020**, *31*, 100892. [CrossRef]
13. Häußler, P.; Emmrich, D.; Müller, O.; Ilzhöfer, B.; Nowicki, L.; Albers, A. Automated topology optimization of flexible components in hybrid finite element multibody systems using ADAMS/Flex and MSC. Construct. In Proceedings of the ADAMS European User's Conference, Berchtesgaden, Germany, 14–15 November 2001; pp. 14–15.
14. Tomlin, M.; Meyer, J. Topology optimization of an additive layer manufactured (ALM) aerospace part. In Proceedings of the 7th Altair CAE Technology Conference, Warwick, UK, 10 May 2011; pp. 1–9.
15. Vayre, B.; Vignat, F.; Villeneuve, F. Designing for additive manufacturing. *Procedia CIrP* **2012**, *3*, 632–637. [CrossRef]
16. Hilpert, E.; Hartung, J.; Risse, S.; Eberhardt, R.; Tünnermann, A. Precision manufacturing of a lightweight mirror body made by selective laser melting. *Precis. Eng.* **2018**, *53*, 310–317. [CrossRef]
17. Jang, J.; Choi, W.C. Error Compensation Using Variable Stiffness in Orbital Grinding. *Int. J. Precis. Eng. Manuf.* **2018**, *19*, 317–323. [CrossRef]
18. Suh, N.P. *Axiomatic Design and Fabrication of Composite Structures: Applications in Robots, Machine Tools, and Automobiles*; Oxford University Press: Oxford, UK, 2005; ISBN 0198039751.
19. Ito, S.; Iijima, D.; Hayashi, A.; Aoyama, H.; Yamanaka, M. Micro turning system: A super small cnc precision lathe for microfactories. In Proceedings of the 3rd International Workshop on Microfactories, Minneapolis, MN, USA, 16–18 September 2002; pp. 37–40.
20. Neugebauer, R.; Wabner, M.; Rentzsch, H.; Ihlenfeldt, S. Structure principles of energy efficient machine tools. *CIRP J. Manuf. Sci. Technol.* **2011**, *4*, 136–147. [CrossRef]
21. Oiwa, T. Coordinate measuring machine using parallel mechanism. In Proceedings of the Proc. 16th IMEKO World Congress, Vienna, Austria, 25–28 September 2000; Volume 8, pp. 211–214.
22. Calignano, F.; Cattano, G.; Manfredi, D. Manufacturing of thin wall structures in AlSi10Mg alloy by laser powder bed fusion through process parameters. *J. Mater. Process. Technol.* **2018**, *255*, 773–783. [CrossRef]
23. Calignano, F.; Galati, M.; Iuliano, L. A metal powder bed fusion process in industry: Qualification considerations. *Machines* **2019**, *7*, 72. [CrossRef]
24. Vayre, B.; Vignat, F.; Villeneuve, F. Metallic additive manufacturing: State-of-the-art review and prospects. *Mech. Ind.* **2012**, *13*, 89–96. [CrossRef]
25. Yang, S.; Zhao, Y.F. Additive manufacturing-enabled design theory and methodology: A critical review. *Int. J. Adv. Manuf. Technol.* **2015**, *80*, 327–342. [CrossRef]
26. Adam, G.A.O.; Zimmer, D. On design for additive manufacturing: Evaluating geometrical limitations. *Rapid Prototyp. J.* **2015**, *21*, 662–670. [CrossRef]

27. Mumtaz, K.A.; Hopkinson, N. Selective laser melting of thin wall parts using pulse shaping. *J. Mater. Process. Technol.* **2010**, *210*, 279–287. [CrossRef]
28. Kumke, M.; Watschke, H.; Vietor, T. A new methodological framework for design for additive manufacturing. *Virtual Phys. Prototyp.* **2016**, *11*, 3–19. [CrossRef]
29. Leary, M.; Merli, L.; Torti, F.; Mazur, M.; Brandt, M. Optimal topology for additive manufacture: A method for enabling additive manufacture of support-free optimal structures. *Mater. Des.* **2014**, *63*, 678–690. [CrossRef]
30. Calignano, F. Design optimization of supports for overhanging structures in aluminum and titanium alloys by selective laser melting. *Mater. Des.* **2014**, *64*, 203–213. [CrossRef]
31. Rickenbacher, L.; Spierings, A.; Wegener, K. An integrated cost-model for selective laser melting (SLM). *Rapid Prototyp. J.* **2013**, *19*, 208–214. [CrossRef]
32. Baumers, M.; Dickens, P.; Tuck, C.; Hague, R. The cost of additive manufacturing: Machine productivity, economies of scale and technology-push. *Technol. Forecast. Soc. Chang.* **2016**, *102*, 193–201. [CrossRef]
33. Moylan, S.; Slotwinski, J.; Cooke, A.; Jurrens, K.; Donmez, M.A. Lessons Learned in Establishing the NIST Metal Additive Manufacturing Laboratory. *NIST Rep.* 2013. [CrossRef]
34. He, G.; Guo, L.; Li, S.; Zhang, D. Simulation and analysis for accuracy predication and adjustment for machine tool assembly process. *Adv. Mech. Eng.* **2017**, *9*, 1687814017734475. [CrossRef]
35. SPEA Automatic Test Equipment Personal comunication. Via Torino, 16, 10088 Volpiano TO, Italy. 2019.
36. Kim, I.Y.; Kwak, B.M. Design space optimization using a numerical design continuation method. *Int. J. Numer. Methods Eng.* **2002**, *53*, 1979–2002. [CrossRef]
37. Lee, S.M.; Han, S.Y. Topology optimization scheme for dynamic stiffness problems using harmony search method. *Int. J. Precis. Eng. Manuf.* **2016**, *17*, 1187–1194. [CrossRef]
38. Gibson, I.; Rosen, D.W.; Stucker, B. Desing for additive manufacturing. In *Additive Manufacturing Technologies: Rapid Prototyping to Drect Digital Manufacturing*; Springer: Boston, MA, USA, 2010; pp. 299–332, ISBN 9781441911193.
39. Hao, L.; Raymond, D.; Strano, G.; Dadbakhsh, S. Enhancing the sustainability of additive manufacturing. In Proceedings of the 5th International Conference on Responsive Manufacturing—Green Manufacturing (ICRM 2010), Ningbo, China, 11–13 January 2010.

crystals

MDPI

Article

Experimental Study of Thermomechanical Processes: Laser Welding and Melting of a Powder Bed

Yassine Saadlaoui *, Julien Sijobert, Maria Doubenskaia, Philippe Bertrand, Eric Feulvarch and Jean-Michel Bergheau

University of Lyon, ENISE, LTDS UMR 5513 CNRS, 58 rue Jean Parot, 42023 Saint-Etienne CEDEX 02, France; julien.sijobert@enise.fr (J.S.); maria.doubenskaia@enise.fr (M.D.); philippe.bertrand@enise.fr (P.B.); eric.feulvarch@enise.fr (E.F.); jean-michel.bergheau@enise.fr (J.-M.B.)

* Correspondence: yassine.saadlaoui@enise.fr

Received: 4 March 2020; Accepted: 23 March 2020; Published: 26 March 2020

Abstract: In this study, an experimental approach was developed to analyze and better understand the laser welding and melting of a powder bed process. Different optical diagnostics tools (high-speed camera, infrared camera, pyrometer, etc.) were applied to measure different physical quantities (molten pool morphology, temperature field, residual stresses, and distortions). As a result, measurements during the laser welding process facilitated the building of a database of experimental results (experimental benchmarks). The study of the melting of a powder bed enabled a better understanding of the physics related to the formation and behavior of the molten pool. These results can be used by researchers to improve and validate numerical simulations of these processes.

Keywords: melting of a powder bed; laser welding; optical diagnostics; molten pool; temperature field; residual stresses

---

1. Introduction

The laser welding process is used widely in international industry. It involves using a laser to assemble various metals (carbon steels, stainless steels, aluminum, and titanium). A laser is a concentrated heat source that can provide a tight and deep weld bead. This process has different advantages over other welding processes: the absence of chamfers and filler material, a high welding speed, a controlled welding depth, a small thermal affected zone, and a low deformation of welded parts. Recently, additive manufacturing processes have begun to occupy an increasingly important place in the industrial world, especially those based on the melting of a powder bed. They consist of producing parts through layer by layer deposition and selective melting of powder layers using a laser. During the melting of a powder bed, different materials can be used: metals, polymers, and ceramics. Metals are by far the most used materials in different industrial environments such as medical, aerospace, automotive, and power.

The laser welding and especially the melting of a powder bed require continuous studying to improve their performance. The experimental approach is often used to study and analyze these thermomechanical processes [1–8]. These studies can be very expensive considering that the processes parameters are numerous [9,10]. In this context, numerical simulation can be an alternative solution. It facilitates the optimization of these parameters and predicting the final characteristics of manufactured parts for a reasonable cost and time.

During the numerical simulation of these thermomechanical processes, several physical phenomena must be considered. These phenomena include the formation of the molten pool, heat transfers, metallurgical transformation, residual stresses, and distortions. Depending on the aims targeted by the simulation, some physical phenomena can be neglected [11]. For example, during numerical modeling that focuses on the formation of the molten pool (thermo-fluid simulation), the mechanical computation (residual stresses and distortions computation) is often overlooked [12–15]. Thermomechanical simulation is generally used to estimate residual stresses and distortions, thus providing very useful information for fatigue lifetime predictions. The thermal computation of this type of simulation is often simplified (the formation of the molten pool is neglected) to reduce the computation time [16–24]. Some studies have attempted to consider most of these physical phenomena. For example, Saadlaoui et al. [25] developed a new strategy to simulate the interaction between the fluid flow in the molten pool and the deformations in the base metal. Therefore, their strategy enables the provision, at the same time, of thermal, fluid, and mechanical results (thermal cycle, molten pool morphology, fluid flow, residual stresses, and deformations). As mentioned above, the numerical simulation can be a useful solution in studying these processes for a reasonable cost and time. However, the numerical simulation is based on different assumptions [11]. For this, a step of validating the numerical models may be necessary. It involves testing the efficiency and reliability of the simulation using experimental or numerical benchmarks. Frequently, researchers prefer to validate their modeling using numerical benchmarks. That is mostly because of the reasonable cost and the accessibility of these benchmarks (contrarily to experimental benchmarks). Here, the problem is that these numerical benchmarks are often simple academic simulations. For example, the sloshing problem was used by Saadlaoui et al. [15] to validate the fluid formulation of the molten pool formation during the laser welding process. It involves tracking the oscillations of a liquid in a container [26]. Therefore, these benchmarks cannot be directly used to validate the results of a laser welding simulation. In this context, several experimental measurements were developed to study and to build a database (experimental benchmarks) of thermomechanical processes [27–32]. These experimental benchmarks can clearly be used to improve and validate numerical simulations.

Some experimental studies have examined the formation of a molten pool during laser welding and the melting of powder bed processes [10,30,33–35]. They have enabled the measuring of the molten pool morphology. This may assist in validating the thermo-fluid simulation of these processes. Gao et al. [36] studied the formation of a molten pool and the keyhole during ND: YAG laser welding of stainless steel. They applied a coaxial visual sensing system to record images of the molten pool. The same system was used by Kim et al. [29] to observe the morphology of the molten pool and keyhole during a remote laser welding. The effect of the welding speed on the molten pool morphology during a high-power laser welding was studied by Li et al. [37]. Lei et al. [32] used a high-speed camera synchronized with an auxiliary laser light to study the effect of laser welding parameters on the molten pool morphology. The same system was used by Trapp et al. [10] to follow the formation of the molten pool during the melting of a powder bed. These authors and others have studied only the formation of the molten pool. In addition, the molten pool images of these studies (especially for the melting of a powder bed) are not always of sufficient quality to determine the molten pool morphology [10,38–42]. Therefore, these studies cannot be used to validate a complete simulation (thermo-fluid-mechanical simulation) of thermomechanical processes. Indeed, information on the temperature field and the residual stresses distribution may be necessary for such validation. These physical quantities were measured separately by some studies [27,31]. Finding an experimental study that measured the molten pool morphology, the temperature field, the residual stresses, and the distortions at the same time is very rare.

In this paper, an experimental approach is proposed to study laser welding. It involves instrumenting this process to measure different physical quantities (molten pool morphology, temperature field, residual stresses, and distortions). These quantities can be used to build a database (experimental benchmark) to validate the numerical simulation of this process. In addition, the formation of the molten pool during the melting of a powder bed process was also studied using the same experimental approach. Here, the aim was to analyze and understand this recent process better.

The paper is organized as follows:

- in Section 2, the proposed experimental setup is presented.
- in Section 3, the methodologies used during the laser welding and the melting of a powder bed are detailed.
- in Section 4, the results of the study of these processes are illustrated and discussed.

2. Experiment Setup

2.1. Proposed Approach

A specific experimental setup (Figure 1) was designed and realized using different instrumentation tools (high-speed camera, infrared camera, etc.).

For laser welding by conduction, the numerical simulation is well advanced and controlled. It enables the consideration of the physical phenomena during this process [11,14,15,43]. For this, the main objective of welding measurements is to acquire the necessary results to validate the numerical simulation of this process. However, obtaining results of all physical quantities using the experimental approach is not always easy. Hence, we will focus on the weld pool morphology, the temperature field, residual stresses, and distortions.

Contrary to laser welding, the numerical simulation of the melting of a powder bed is not sufficiently advanced [44–47]. Thus, the first interest of the experimental study was to understand the physical phenomena involved better. This can help to improve the numerical simulation of this process. We note that the same experimental setup was used to study both processes (laser welding and melting of a powder bed).

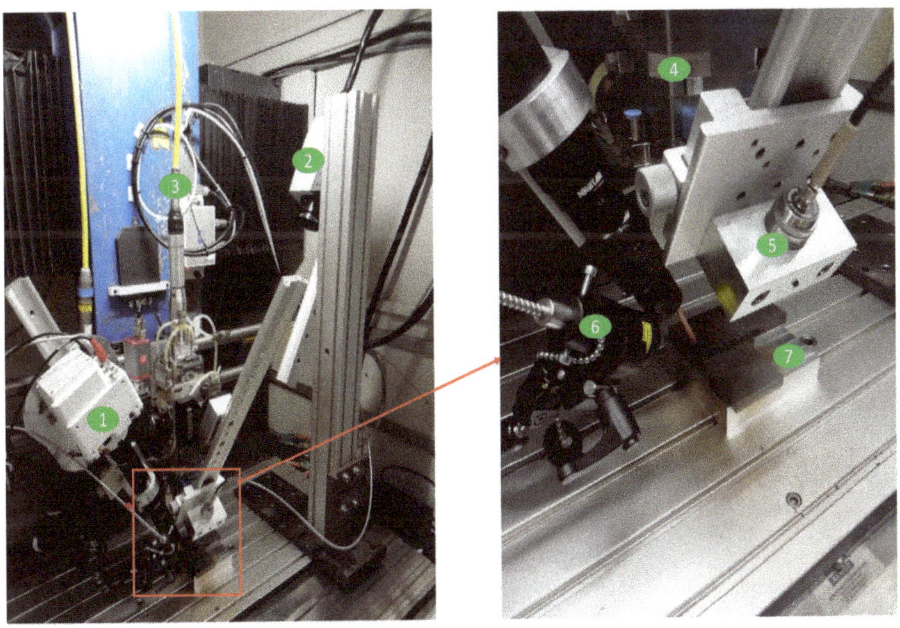

Figure 1. Experiment setup: 1—High-speed camera; 2—Infrared camera; 3—Laser fiber; 4—Laser head; 5—Pyrometer; 6—Auxiliary laser light; 7—Samples support.

2.2. Experimental Equipment and Samples

2.2.1. Machine and Laser

A TRUMPF LASMA 1054 CNC machine (3 axis) was used (Figure 1). A PRECITEC YC52 laser head was powered by a LASERLINE LDM 3000-60 diode laser with a theoretical power of 3 kW and a wavelength of 930 nm. A real laser power equal to 2.5 kW was measured using a wattmeter. The beam analysis of this laser was carried out for the maximum power (2.5 kW) and the results are given in Figure 2.

This analysis enables us to position the sample in the focal plane. Initially, we decided that the upper surface of the sample is merged with this plane (Figure 3a). This configuration causes the formation of a keyhole that is not often considered during numerical simulations [15]. For this reason, the laser beam was defocused by 5 mm (Figure 3b) to conduct a laser welding by conduction. This second configuration give a laser beam diameter of 2.4 mm. The energy distributions of the laser beam of these two configurations are given by Figure 3c,d. The formation of the molten pool during the two welding configurations is presented in Figure 4.

2.2.2. High-Speed Camera and Auxiliary Laser Light

A Phantom VEO 710 high-speed camera with a resolution of 1280 × 800 pixels was used. With a maximal rate of 7400 fps, this camera model is perfect to track the formation of the molten pool. Phantom Camera Control (PCC) software was used to configure, control, process, and download images. A filter with a wavelength of 810 nm was installed on the camera lens (NAVITAR ZM 6000-II, 12 mm FF). The camera was also synchronized with an auxiliary laser light (CAVILUX diode laser with a wavelength of 810 nm and a maximal power of 500 W). This facilitated the capturing of clear images of the molten pool surface (without the light of the laser power). As shown in Figure 1, the camera was attached to the laser head to monitor the molten pool movement. The working distance (the distance between the camera lens and the tip of the power laser) was approximately 100 mm and the tilt angle of the camera from the vertical was approximately 30°. We note that a reference sample was used to determine the scale of the camera images.

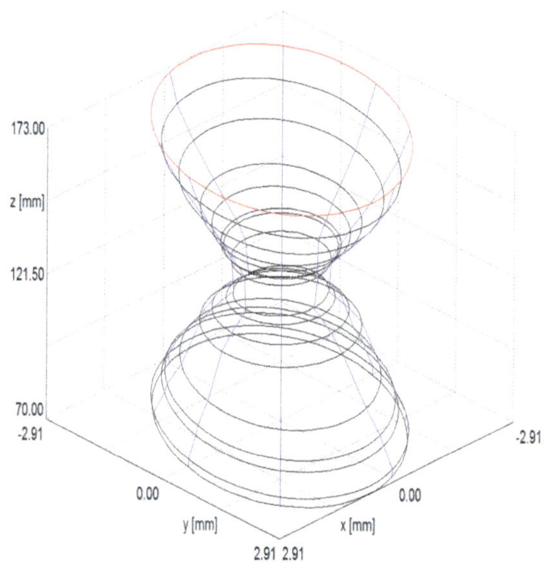

Figure 2. Laser beam analysis for maximal power of 2.5 kW.

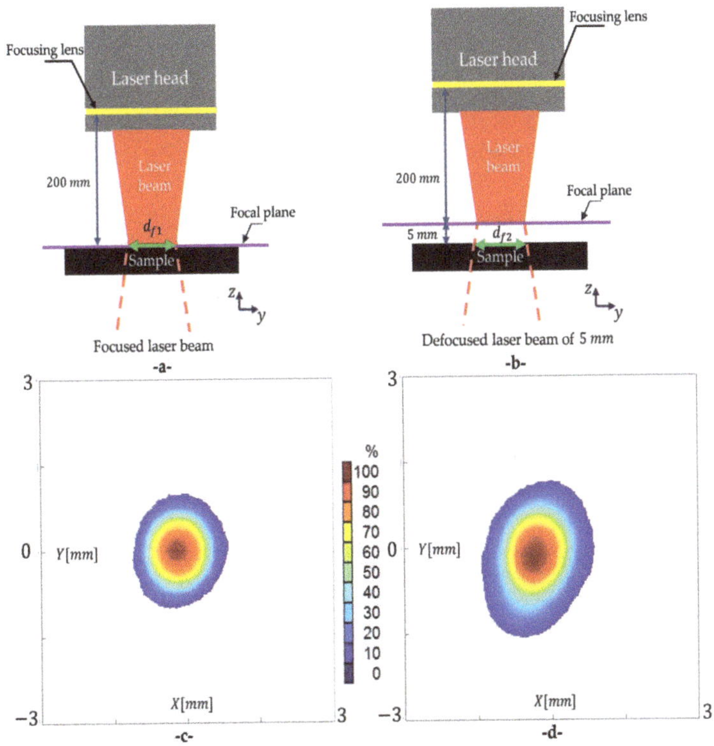

Figure 3. (**a**) Focused laser beam; (**b**) Defocused laser beam of 5 mm; (**c**) Energy distribution of focused laser beam; (**d**) Energy distribution of defocused laser beam.

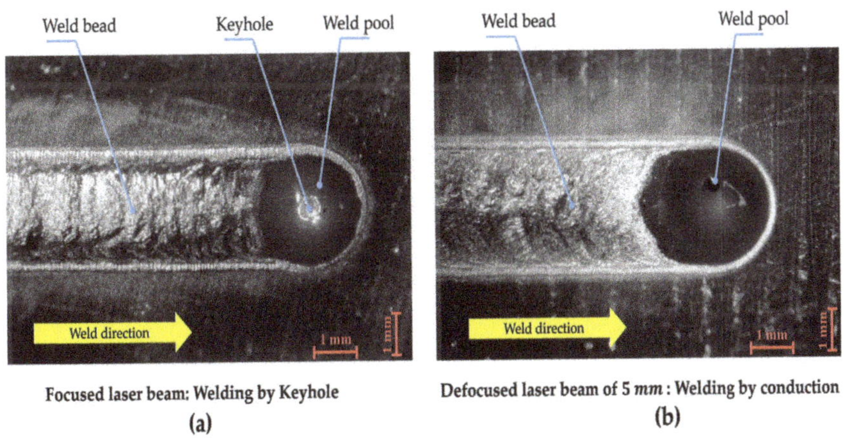

Figure 4. Two welding configuration (images of high-speed camera): (**a**) welding by keyhole; (**b**) welding by conduction (P = 2.5 KW, v = 600 mm.min^{-1}).

2.2.3. Infrared Camera and Pyrometer

A FLIR Phoenix MWIR infrared camera with a resolution of 320 × 256 pixels was used to determine the temperature distribution in the weld bead. As Figure 1 shows, the camera was attached to the machine table. Its working distance and tilt angle were in the order of 695 mm and 29°, respectively. With this configuration, the camera facilitated the capturing of the temperature distribution throughout the weld bead. An attenuator filter was installed on the camera lens to capture temperatures approximately 1600 °C. An integration time of 0.01 ms and acquisition time of 10 µs were chosen.

We note that the infrared camera was not calibrated, so it could not provide temperature values. For this, a KLEIBER LVO 25 S-7 monochromatic pyrometer was used to measure the temperature in the weld bead. It can measure temperatures between 350 and 3500 °C. A specific mounting was used to install the pyrometer on the laser head (Figure 1). This specific mounting expedited moving the pyrometer and measuring the temperature in four positions (P1, P2, P3, and P4) on the weld bead (Figure 5). A working distance of approximately 123 mm and a length of the pyrometer body of 61 mm enabled the measuring of the temperature with a spot diameter of 0.9 mm. We note that the emissivity of the material was assumed to be 1. Therefore, the temperature given by the pyrometer was the brightness temperature.

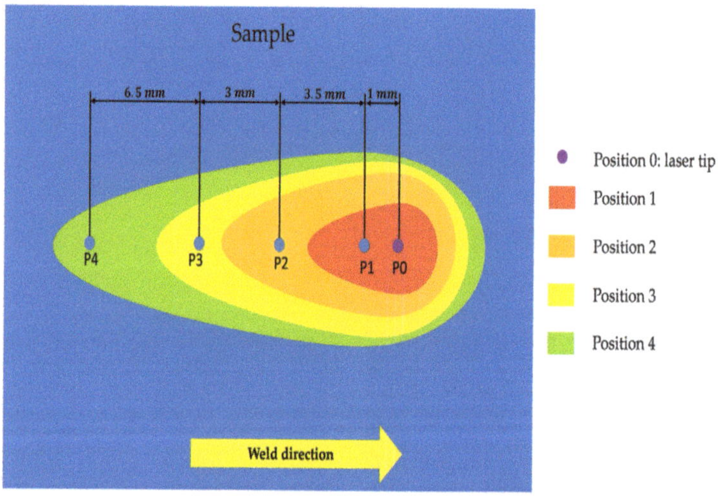

Figure 5. Measurement points of the pyrometer.

2.2.4. Optical Microscope, DRC Machine, and 3D Measuring Arm

The length and width of the weld pool can be measured using the images of the high-speed camera. However, to determine its depth and height, the use of another approach is necessary. It involves making samples by transversely cutting the welded samples. These samples must then undergo a polishing step. Subsequently, a chemical attack must be conducted to distinguish the different zones: weld bead, heat affected zone (HAZ), and the base metal. The final step is to visualize the samples using an optical microscope.

A PROTO DRX machine (with a collimator diameter of 1 mm) was used to measure residual stresses. The detector uses two orientations, 0° and 90°, to determine the axial and longitudinal residual stresses. For each of these two orientations, the scanned angles are −25°; −16.66°; −8.33°; 0°; 8.33°; 16.66°, and 25°.

A FARO GAGE 3D measuring arm (with a precision of 0.025 mm) was used to measure the flatness of the samples after welding. This measuring enabled us to determine the vertical displacements U_z of the welded samples (Figure 6).

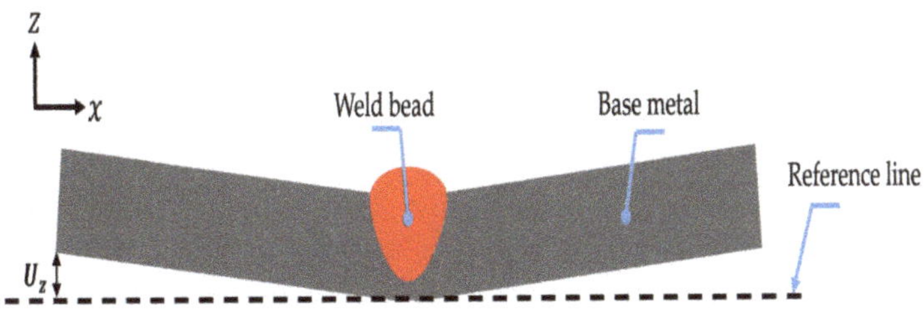

Figure 6. Vertical displacement U_z of welded sample: transverse plane.

2.2.5. Samples Preparation

For the laser welding, 316L stainless steel plates (austenitic steel) from hot rolling were machined to obtain nine samples with dimensions of 100 mm × 100 mm × 10 mm. The chemical composition of this material is given in Table 1. These samples were subjected to a relaxation heat treatment at 610 °C for 2.5 h with slow cooling in a neutral atmosphere (nitrogen). This heat treatment facilitates the reduction of the residual stresses related to the machining and rolling processes. Thus, this enables avoiding considering the initial residual stresses during the numerical simulation of the laser welding process. The first measurement of stresses (after machining of samples and before the heat treatment) indicated values in the order of 700 MPa while values of approximately 100 MPa were observed after this heat treatment. Samples in 15CDV6 low alloy steel (bainitic steel) were also machined (with the same dimensions). The aim here was to study the formation of the weld pool of two materials that do not have the same thermophysical properties.

Table 1. Chemical composition of 316L stainless steel in percent.

C	Mn	P	S	Si	Cr	Ni	Mo	N
0.022	1846	0.039	0.0004	0.319	16.97	20.01	2.01	0.0368

The melting of a powder bed involved filling a hollow 316L substrate of dimensions 60 mm × 60 mm × 10 mm (Figure 7) by a layer of 316L powder. Subsequently, the power laser facilitated the formation of a bead by melting a zone of this layer (thickness of 2 mm). Two types of powder were examined to study the influence of the size and granulometry of the powder on the formation of the molten pool: a spherical powder with a diameter between 20 and 50 μm, and another irregular one with a size between 100 and 150 μm (Figure 8).

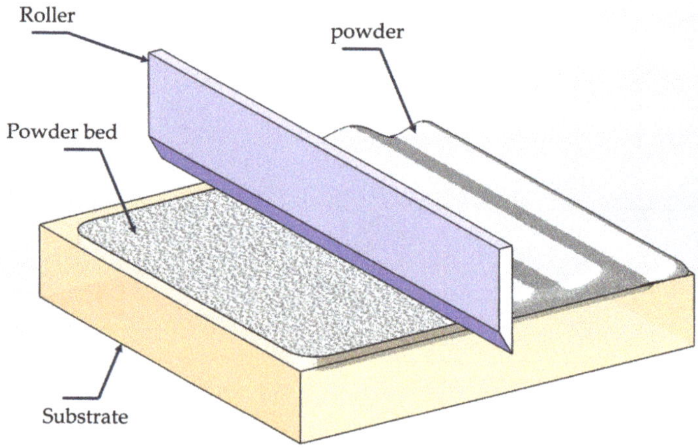

Figure 7. Preparation of the powder layer.

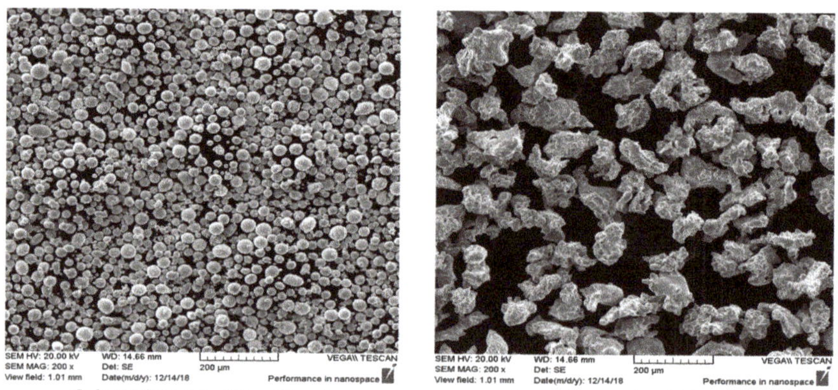

Spherical powder $(20 - 50\ \mu m)$

(a)

Irregular powder $(100 - 150\ \mu m)$

(b)

Figure 8. Granulometry of 316L powder: (**a**) spherical powder (20–50 µm); (**b**) irregular powder (100–150 µm).

3. Methodology

3.1. Laser Welding

Laser welding tests involved creating a bead of length 90 mm on each sample. A maximal power of laser of approximately 2.5 kW was set. Welding speeds of 600, 1600, and 2600 mm.min^{-1} were tested. Three 316L samples were studied for each speed. 15CDV6 samples were also examined to study the weld pool formation according to the welded material. A comparison of the weld pool morphology of the two materials (316L and 15CDV6) was conducted.

As noted earlier, the aim here was to create a database (experimental benchmark) that enables the validating of numerical simulations of laser welding. To achieve this, we were interested in the weld pool morphology, temperature evolution, residual stresses, and distortions of the 316L samples. The length and width of the weld pool were measured from the images of the high-speed camera. However, to determine its depth and height, the approach proposed in Section 2.2.4 must be used. The infrared camera enabled us to obtain the temperature distribution. At the same time, the pyrometer enabled us to measure the brightness temperature for each color given by the infrared camera (points P1, P2, P3, and P4: Figure 5). Residual stresses related to the welding process were also measured at three points (1, 2, and 3) on 316L samples (Figure 9). The welding conditions and measurements performed for the two materials are given in Table 2.

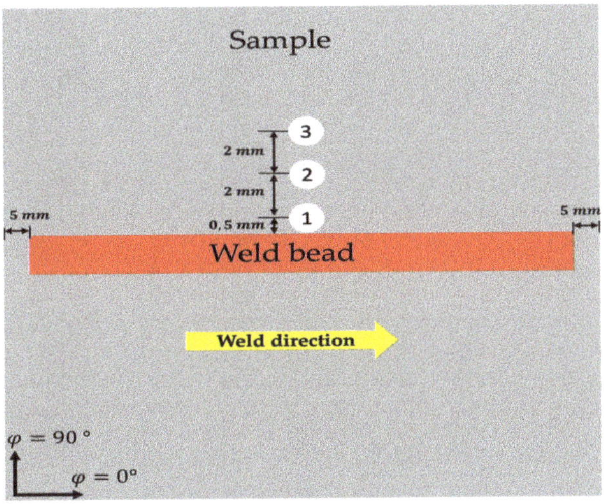

Figure 9. Measurement points of residual stress.

Table 2. Machine parameters and measured quantities for both materials (316L and 15CDV6).

Material	Power [kW]	Speed [mm.min^{-1}]	Molten Pool Morphology	Temperature	Residual Stresses	Distortions
316L	2.5	600, 1600, and 2600	measured	measured	measured	measured
15CDV6	2.5	600, 1600, and 2600	measured	not measured	not measured	not measured

3.2. *Melting of a Powder Bed*

A single bead of 50 mm length was created by melting a zone of the powder layer. Similar to laser welding, the maximum power of the laser was used (2.5 kW). Scanning speeds (speed of laser movement) between 400 and 3600 mm.min^{-1} were examined. Three beads were created for each speed.

4. Results and Discussion

In this section, the results of the laser welding process (molten pool morphology, temperature field, residual stresses, and distortions) are given and discussed. The authors are interested in the effect of the welding speed on these results. The results (formation of the molten pool) of the melting of a powder bed are then presented.

4.1. *Laser Welding*

4.1.1. Weld Pool Morphology

Figure 10 shows the microscopic images of a cross-section of 316L and 15CDV6 weld beads for a speed of 1600 mm.min^{-1}. These images enable us to determine the depth, height, and width of the weld bead. The length of the molten pool can be measured from images of the high-speed camera (the width can also be measured from these images: Figure 11). Tables 3 and 4 illustrate respectively the dimensions of the 316L and 15CDV6 weld beads for three speeds (600, 1600 and 2600 mm.min^{-1}). Figure 12 shows the high-speed camera images of the weld pool of the two materials for a welding speed equal to 1600 mm.min^{-1}.

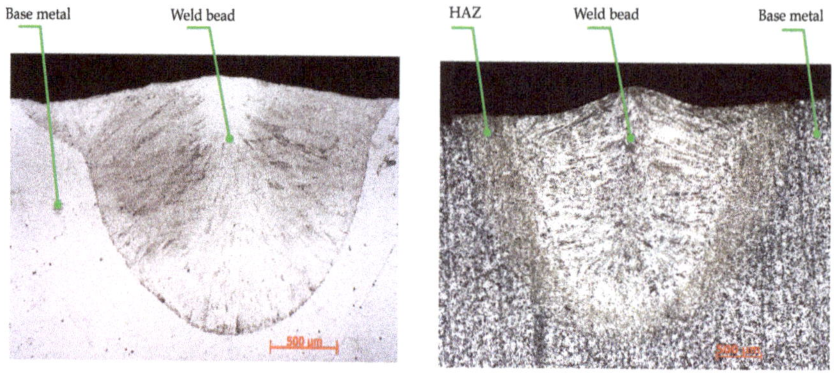

316L
(a)

15CDV6
(b)

Figure 10. Cross-section of the weld bead (microscopic images, v = 1600 mm.min^{-1}): (**a**) 316L; (**b**) 15CDV6.

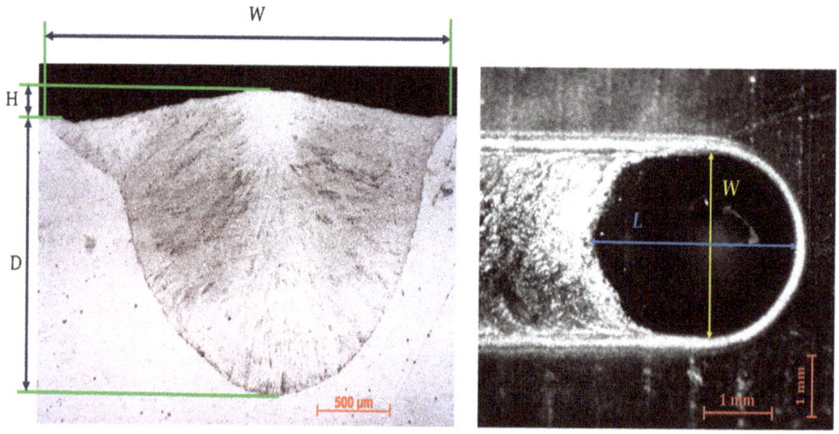

Figure 11. Morphology of the weld pool (316L): 1600 mm.min^{-1}.

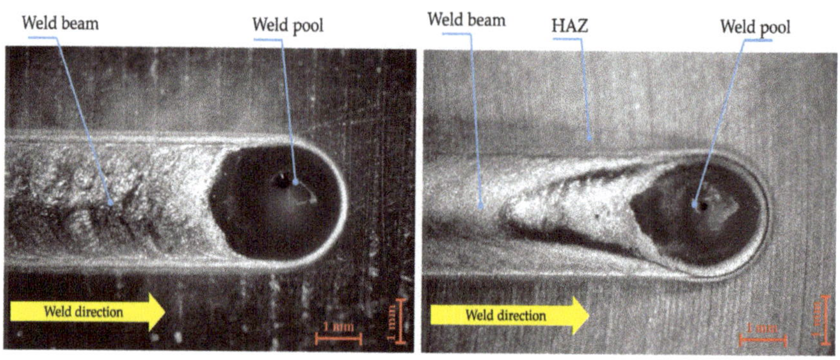

316L
(a)

15CDV6
(b)

Figure 12. Morphology of the weld pools of both materials (1600 mm.min^{-1}): (**a**) 316L; (**b**) 15CDV6.

Table 3. Morphology of the weld pool (316L).

Speed [mm.min^{-1}]	*L* [mm]	*W* [mm]	*D* [mm]	*H* [mm]
600	2.85 ± 0.2	2.97 ± 0.1	2.12 ± 0.1	0.31 ± 0.02
1600	2.98 ± 0.2	2.78 ± 0.1	1.29 ± 0.1	0.15 ± 0.02
2600	3.1 ± 0.2	2.6 ± 0.1	0.99 ± 0.1	0.12 ± 0.02

Table 4. Morphology of the weld pool (15CDV6).

Speed [mm.min^{-1}]	*L* [mm]	*W* [mm]	*D* [mm]	*H* [mm]
600	5.27 ± 0.2	2.82 ± 0.1	2.14 ± 0.1	0.25 ± 0.02
1600	5.56 ± 0.2	2.15 ± 0.1	1.16 ± 0.1	0.1 ± 0.02
2600	5.9 ± 0.2	2.06 ± 0.1	0.95 ± 0.1	0.09 ± 0.02

The images captured by the optical microscope and the high-speed camera enabled us to build a database of 316L (Table 3) and 15CDV6 (Table 4) weld pool morphologies. Tables 3 and 4 indicate a significant effect of the welding speed on the weld pool morphology. As expected, the weld pool is longer and shallower for the speed of 2600 mm.min^{-1}. In Figure 12, the weld pool, HAZ, and base metal can be clearly distinguished. Because 316L steel is austenitic, it does not undergo a major metallurgical transformation during welding (not HAZ). However, the HAZ is very remarkable for the 15CDV6 material (bainitic steel). Figure 12 also shows that the weld pool morphology is sensitive to the welded material. For example, using the same welding conditions, the 15CDV6 weld pool is longer compared to the 316L weld pool (Tables 3 and 4). This may be justified by the fact that the two materials do not have the same thermophysical properties (thermal conductivity, viscosity, surface tension, etc.).

4.1.2. Temperature Field

The temperature distribution during the laser welding of the 316L samples is shown in Figure 13. We note that an arbitrary unit (A.U.) was used in Figure 13 because the infrared camera was not calibrated (the scale does not represent temperature values). Figure 14 illustrates the evolution of the brightness temperature in the melted zone according to welding speed. The evolution of this temperature in different positions of the weld bead is given in Figure 15 (v = 1600 mm.min^{-1}).

The images of the temperature distribution captured by the infrared camera (Figure 13) show that the welding speed affects the isotherms distribution (and thus the morphology of the weld pool). The pyrometer enabled us to determine the brightness temperature values (Figure 15) corresponding to the different colors captured by the infrared camera. This brightness temperature in the weld pool is not very sensitive to the welding speed (Figure 14).

We note that the true temperature restoration is the key problem of pyrometry because of the unknown emissivity and its possible variation during measurement. Different methods can be applied but no universal solution is found [27,31]. The brightness temperature of a non-blackbody target is defined as the temperature of a blackbody with the same monochromatic luminance. The relation between the two temperatures is presented in [48]. The true temperature can be measured during a future work by using other experimental approaches (bichromatic pyrometer, thermocouple, etc.).

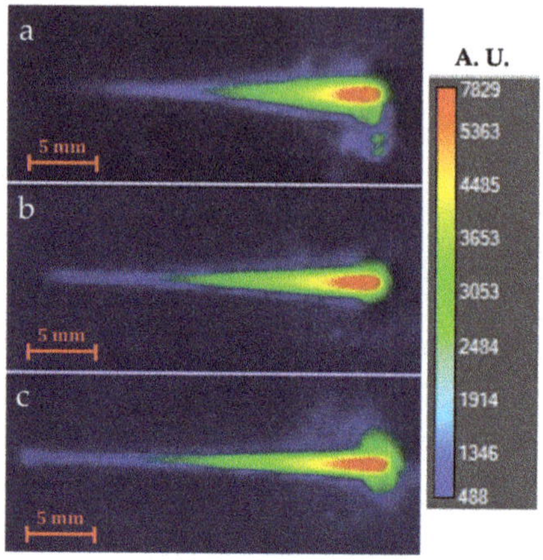

Figure 13. Temperature field given by infrared camera (316L): (**a**) v = 600 mm.min^{-1}; (**b**) v = 1600 mm.min^{-1}; (**c**) v = 2600 mm.min^{-1}.

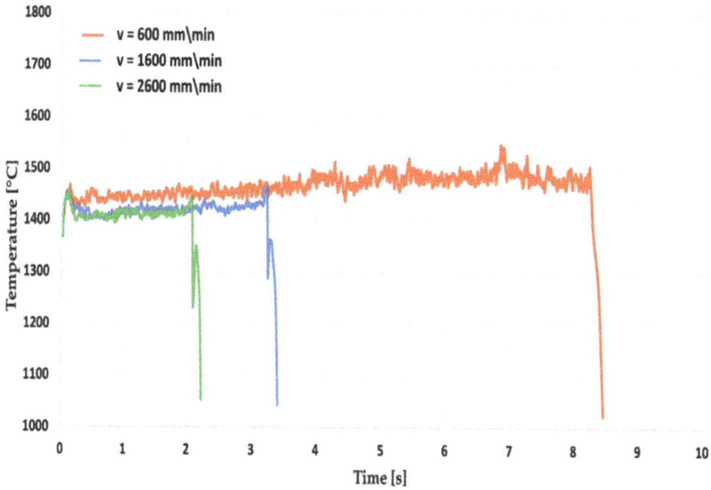

Figure 14. Effect of the welding speed on the brightness temperature evolution (316L).

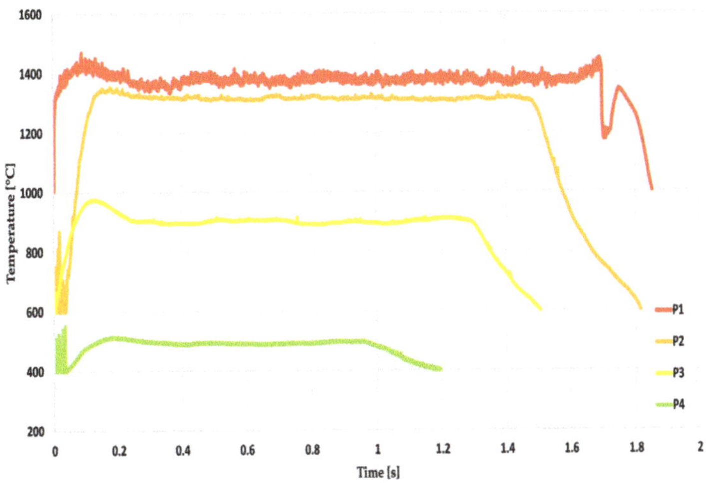

Figure 15. Measurement of the brightness temperature in different positions on the weld bead for v = 1600 mm.min^{-1}: P1, P2, P3, and P4 (316L).

4.1.3. Residual Stresses and Distortions

Axial and longitudinal residual stresses were measured on the 316L samples after welding using the DRX machine. As Figure 9 shows, three measurement points were chosen. We specify that the first point is tangent to the weld bead. Table 5 summarizes the results of these stresses for speeds of 600, 1600, and 2600 mm.min^{-1}.

The vertical displacements of the welded samples were measured using the measuring arm. Vertical displacements of U_z of 0.411, 0.174, and 0.091 mm were measured respectively for the welding speeds of 600, 1600, and 2600 mm.min^{-1}.

Table 5. Residual stresses of 316L samples (points 1, 2, and 3) after welding.

	Stresses [MPa] (ϕ = 0°)			Stresses [MPa] (ϕ = 90°)		
Speed [mm.min^{-1}]	**Point 1**	**Point 2**	**Point 3**	**Point 1**	**Point 2**	**Point 3**
600	688 ± 30	359 ± 11	69 ± 19	−18 ± 30	312 ± 62	323 ± 14
1600	591 ± 25	151 ± 12	73 ± 20	259 ± 30	545 ± 62	405 ± 14
2600	343 ± 22	99 ± 30	75 ± 12	249 ± 30	507 ± 17	278 ± 13

Residual stresses in the order of 100 MPa were observed before the welding of the 316L samples. These stresses increased after welding to reach approximately 688 MPa for a speed of 600 mm.min^{-1}. This indicates that the residual stresses related to the laser welding process can be very important. As shown in Table 5, these stresses are very sensitive to the welding speed. For example, the longitudinal residual stresses decrease when the speed increases. The vertical displacements also decrease when the welding speed increases ($U_z = 0.411$ for v = 600 mm.min^{-1} and $U_z = 0.091$ for v = 2600 mm.min^{-1}). This can be justified by the effect of the welding speed on the molten pool morphology and thermal cycle.

The results of the laser welding process enabled us to conduct an experimental benchmark (molten pool morphology, temperature field, residual stresses, and distortions). This benchmark can assist in validating the different types of numerical simulations of this process. For example, the molten pool morphology and temperature field results can be used to validate thermo-fluid simulations. The molten pool morphology and especially the free surface shape can also indicate that the tension surface gradient is positive (for used welding conditions). The mechanical results of the numerical simulation can be validated using the residual stresses and distortion results.

4.2. Morphology of the Molten Pool: Melting of a Powder Bed

Figure 16 shows the formation of the molten pool during the melting of the spherical powder. The effect of the scanning speed on the molten pool morphology is given in Figures 17 and 18. The influence of the types of powder on the formation of the molten pool is presented in Figures 19 and 20. Tables 6 and 7 summarize the dimensions of the respective beads of the spherical and irregular powders.

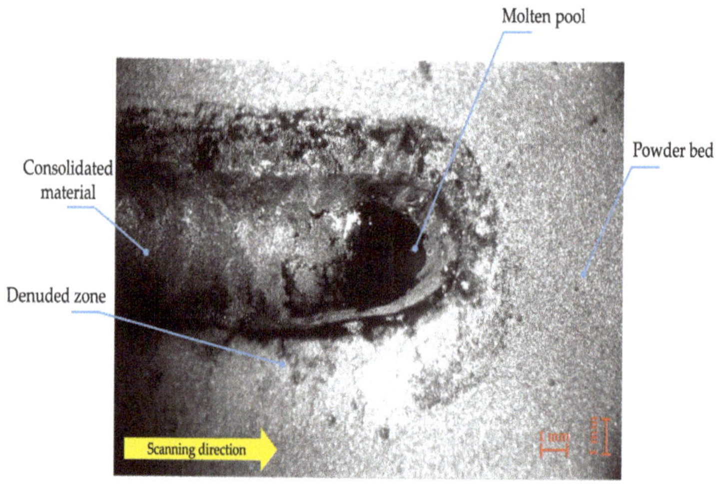

Figure 16. Formation of the molten pool during melting of a 316L spherical powder bed (v = 1600 mm.min^{-1}).

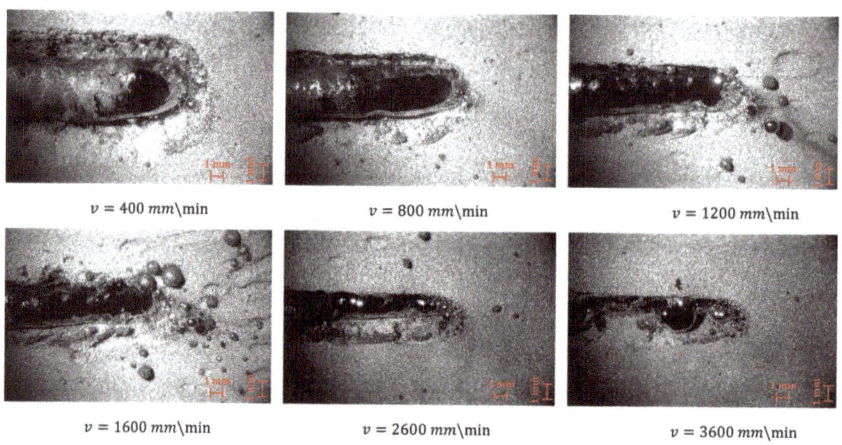

Figure 17. Effect of scanning speed on molten pool morphology (spherical powder): high-speed camera images.

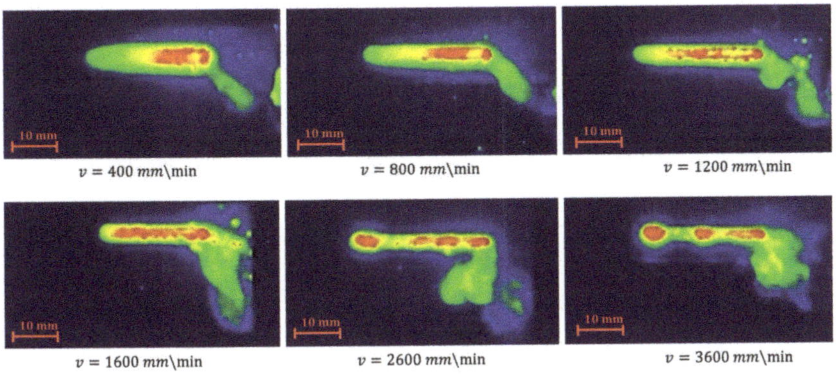

Figure 18. Effect of scanning speed on molten pool morphology (spherical powder): infrared camera images.

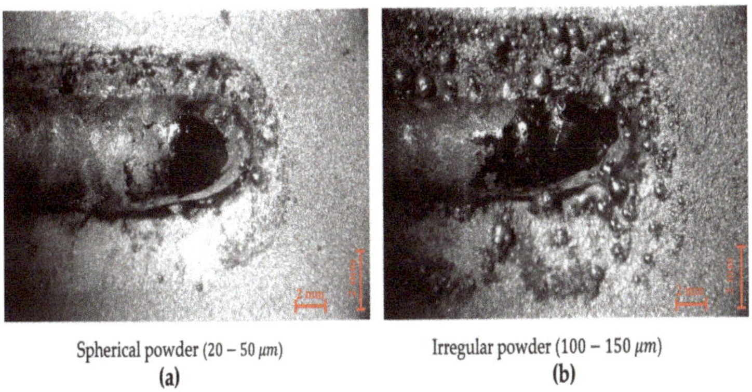

Figure 19. Effect of powder distribution on molten pool morphology (high-speed camera images, v = 400 mm.min^{-1}): (**a**) Spherical powder; (**b**) Irregular powder.

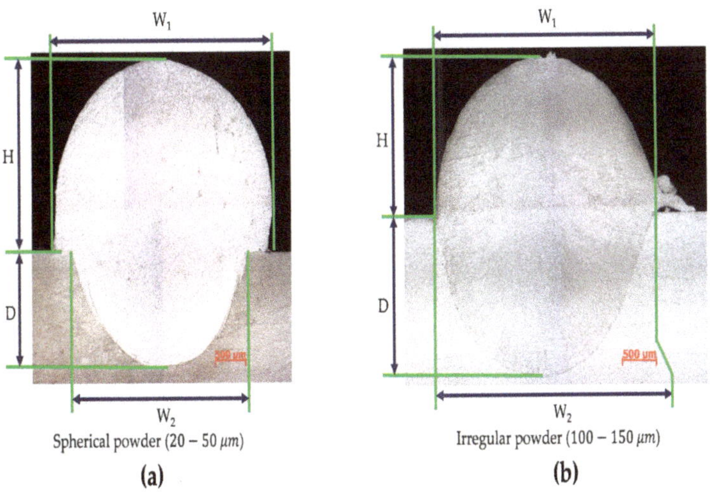

Figure 20. Molten pool morphology (microscopic images, v = 400 mm.min^{-1}): (**a**) Spherical powder; (**b**) Irregular powder.

Table 6. Molten pool morphology of spherical powder (20–50 μm).

Speed [mm.min^{-1}]	W_1 [mm]	W_2 [mm]	D [mm]	H [mm]
400	3.93 ± 0.1	3.17 ± 0.1	1.86 ± 0.1	2.47 ± 0.1
600	3.91 ± 0.1	2.96 ± 0.1	1.61 ± 0.1	2.24 ± 0.1
800	3.53 ± 0.1	2.83 ± 0.1	1.39 ± 0.1	2.22 ± 0.1
1000	3.25 ± 0.1	2.62 ± 0.1	1.3 ± 0.1	2.05 ± 0.1
1200	3 ± 0.1	2.41 ± 0.1	1.09 ± 0.1	1.88 ± 0.1
1400	2.8 ± 0.1	2.21 ± 0.1	1.01 ± 0.1	1.82 ± 0.1

Table 7. Molten pool morphology of irregular powder (100–150 μm).

Speed [mm.min^{-1}]	W_1 [mm]	W_2 [mm]	D [mm]	H [mm]
400	3.77 ± 0.1	3.77 ± 0.1	2.42 ± 0.1	1.99 ± 0.1
600	3.46 ± 0.1	3.46 ± 0.1	1.97 ± 0.1	1.7 ± 0.1
800	3.23 ± 0.1	3.23 ± 0.1	1.7 ± 0.1	1.63 ± 0.1

Figure 16 shows the formation of the molten pool during the melting of a spherical powder layer. In this figure, the molten pool, the bead (consolidated material), and the unfused powder can be distinguished. A denuded zone can also be distinguished; it appears between the bead and the unfused powder. The videos captured by the high-speed camera can explain the causes of the appearance of this zone. Indeed, the molten pool absorbs the powder around it. This absorption can be justified by the mechanical flow of the powder which moves it towards the molten pool. This flow is probably less important for low layer thicknesses.

Figure 17 shows the effect of the scanning speed on the formation of the molten pool. The latter is longer and narrower for the higher velocities. The "balling" phenomenon appears from a speed of 3600 mm.min^{-1} (Figures 17 and 18). Figure 19 shows the molten pool formation of both types of powder (spherical and irregular powders). We note that the type of powder has a direct effect on the formation and morphology of the bead. This can be justified by the influence of the size and the granulometry of the powder on its mechanical flow. A standardized flowability test (ASTM) was used to measure the rate of this flow. It involves timing the flow of 50 *g* of powder through a hole with a diameter of 2.54 mm. The spherical powder flowed after 16 *seconds* while 38 *seconds* was required for the flow of the irregular powder. These results confirm that the flowability of the spherical powder is greater than that of the irregular powder.

The cross-section of the beads of the spherical and irregular powders are given in Figure 20. We observed that the morphology of the bead is sensitive to the types of powder. For example, gaps of 12 and 24% were observed respectively for the width (L1) and height (H) for the speed of 600 mm.min^{-1}. This difference can also be justified by the flowability of the two powders. We have observed that the height of the spherical powder bead is greater than the initial thickness of the powder layer (2 mm) for speeds of 400, 600, 800, and 1000 mm.min^{-1}. This can be explained by the flowability of the spherical powder which facilitates its absorption by the molten pool. This absorption facilitates the increase in the volume of the bead. For these velocities, the phenomenon of shrinkage of the layer powder (shrinkage related to the variation of the density between the powder bed and the molten pool) is, therefore, less visible. However, it begins to be observed from a speed of 1200 mm.min^{-1}. Indeed, the powder quantity absorbed is less important for high velocities. The flowability of the irregular powder and therefore its absorption by the molten pool are less important. For this reason, the height (H) of the irregular powder bead is less than 2 mm, even for low velocities.

The study of the melting of a powder bed enabled us to analyze and understand this process better. As the first remark, we have observed that the molten pool absorbs the unfused powder around it. This phenomenon was very visible especially for the spherical powder, which has a significant flowability. The absorption of the powder by the molten pool leads to the formation of a denuded zone between the bead and the unfused powder. It also facilitates the increase in the bead volume, especially for the spherical powder and at low velocities. Therefore, the shrinkage phenomenon simulated by several studies [46,49] cannot always occur during the melting of a powder bed. However, considering the material addition (related to the mechanical flow of the powder around the molten pool) during the numerical simulation may be necessary according to manufacturing conditions. Considering the fluid flows in the molten pool (related to the absorption of the powder by the molten pool) will also be necessary. The formation of an interface between the molten zone and the unfused powder can also be simulated.

5. Conclusions

An experimental setup was proposed to study the laser welding and melting of powder bed processes. The first aim of this setup was to analyze and understand these two processes better. The second aim was to build an experimental database (experimental benchmark) to validate the different types of numerical simulations of laser welding. The study involved instrumenting a laser welding machine to measure different quantities (molten pool morphology, temperature field, residual stresses, and distortions). Different instrumentation tools (high-speed camera, infrared camera, pyrometer, etc.) were used to achieve this.

In the first step, tests of laser welding were carried out on 316L samples. The effect of the welding speed on the molten pool morphology, temperature field, residual stresses, and distortions was studied. The measured quantities were very sensitive to this speed. For example, the weld pool was longer and shallower for high velocities. The longitudinal residual stresses and distortions decreased when the speed increased. The results of these different quantities will enable researchers to validate their numerical simulations of this process.

In the second step, tests of the melting of a powder bed were carried out using a laser welding machine. The idea was to fill a hollow substrate with a powder layer and then melt it using a laser. The aim of these tests was to understand this process better. As a result, the bead dimensions of two types of powder (spherical and irregular powders) were measured. These results, and especially the height of the bead, depicted that the shrinkage of the powder layer after melting cannot always be visible. Indeed, the height of the bead of the spherical powder exceeded the initial thickness of the powder layer for low velocities. This can be explained by the high-speed camera videos. They indicated that the molten pool absorbed the powder around it. This absorption facilitates the increase in the bead volume and creating a denuded zone between the bead and the unfused powder. The shrinkage of the powder layer was easily visualized with the irregular powder even for low scanning velocities. This can be explained by its low flowability, which delays its absorption by the molten pool (less material addition). The results also indicated that the bead morphology is sensitive to scanning velocities.

Finally, this study enabled us to understand the physics during laser welding and especially during the melting of a powder bed process better. It also enabled us to build a database of experimental results (experimental benchmark) of laser welding. This can assist in improving and validating the numerical simulation of these processes.

Author Contributions: Conceptualization, Y.S., J.S. and J.-M.B.; methodology, Y.S., M.D., E.F. and J.-M.B.; software, Y.S. and J.S.; validation, Y.S., M.D., E.F. and J.-M.B.; formal analysis, Y.S., J.S. and J.-M.B.; investigation, Y.S.; resources, P.B., E.F. and J.-M.B.; data curation, Y.S. and J.S.; writing–original draft preparation, Y.S.; writing–review and editing, J.S., M.D., P.B. and J.-M.B.; visualization, Y.S., P.B. and J.-M.B.; supervision, J.S., M.D. and J.-M.B.; project administration, J.-M.B.; funding acquisition, P.B., E.F. and J.-M.B. All authors have read and agreed to the published version of the manuscript.

Funding: This research received no external funding.

Acknowledgments: The authors would like to thank ESI-GROUP and MELTED project for funding this study.

Conflicts of Interest: The authors declare no conflict of interest.

References

1. Liu, Q.S.; Mahdavian, S.; Aswin, D.; Ding, S. Experimental study of temperature and clamping force during Nd:YAG laser butt welding. *Opt. Laser Technol.* **2009**, *41*, 794–799. [CrossRef]

2. Li, L.; Peng, G.; Wang, J.; Gong, J.; Li, H. Experimental study on weld formation of Inconel 718 with fiber laser welding under reduced ambient pressure. *Vacuum* **2018**, *151*, 140–147. [CrossRef]
3. Pankaj, P.; Tiwari, A.; Bhadra, R.; Biswas, P. Experimental investigation on CO_2 laser butt welding of AISI 304 stainless steel and mild steel thin sheets. *Opt. Laser Technol.* **2019**, *119*, 105633. [CrossRef]
4. Li, J.; Sun, Q.; Kang, K.; Zhen, Z.; Liu, Y.; Feng, J. Process stability and parameters optimization of narrow-gap laser vertical welding with hot wire for thick stainless steel in nuclear power plant. *Opt. Laser Technol.* **2019**, *123*, 105921. [CrossRef]
5. Pal, S.; Lojen, G.; Kokol, V.; Drstvenšek, I. Reducing porosity at the starting layers above supporting bars of the parts made by Selective Laser Melting. *Powder Technol.* **2019**, *355*, 268–277. [CrossRef]
6. Zhang, B.; Xiu, M.; Tan, Y.T.; Wei, J.; Wang, P. Pitting corrosion of SLM Inconel 718 sample under surface and heat treatments. *Appl. Surf. Sci.* **2019**, *490*, 556–567. [CrossRef]
7. Gao, X.; Liu, Z.; Li, J.; Liu, E.; Yue, C.; Zhao, K.; Yang, G. Selective laser melting of CuZr-based metallic glass composites. *Mater. Lett.* **2020**, *259*, 126724. [CrossRef]
8. Chen, H.; Zi, X.; Han, Y.; Dong, J.; Liu, S.; Chen, C. Microstructure and mechanical properties of additive manufactured W-Ni-Fe-Co composite produced by selective laser melting. *Int. J. Refract. Met. Hard Mater.* **2020**, *86*, 105111. [CrossRef]
9. Sun, J.; Yang, Y.; Wang, D. Parametric optimization of selective laser melting for forming Ti6Al4V samples by Taguchi method. *Opt. Laser Technol.* **2013**, *49*, 118–124. [CrossRef]
10. Trapp, J.; Rubenchik, A.M.; Guss, G.; Matthews, M.J. In situ absorptivity measurements of metallic powders during laser powder-bed fusion additive manufacturing. *Appl. Mater. Today* **2017**, *9*, 341–349. [CrossRef]
11. Dal, M.; Fabbro, R. [INVITED] An overview of the state of art in laser welding simulation. *Opt. Laser Technol.* **2016**, *78*, 2–14. [CrossRef]
12. Ko, S.H.; Yoo, C.D.; Farson, D.F.; Choi, S.K. Mathematical modeling of the dynamic behavior of gas tungsten arc weld pools. *Metall. Mater. Trans. B* **2000**, *31*, 1465–1473. [CrossRef]
13. Khairallah, S.; Anderson, A.; Rubenchik, A. Laser powder-bed fusion additive manufacturing: Effects of main physical processes on dynamical melt flow and pore formation from mesoscopic powder simulation. *Acta Mater.* **2015**.
14. Saadlaoui, Y.; Sallem, H.; Feulvarch, E.; Sayet, T.; Leblond, J.b.; Bergheau, J.M. Influence des écoulements de la matière sur l'évolution des contraintes résiduelles durant le procédé de soudage. In Proceedings of the 13e Colloque National en Calcul des Structures CSMA 2017, Giens, Var, France, 15–19 May 2017.
15. Saadlaoui, Y.; Feulvarch, E.; Delache, A.; Leblond, J.B.; Bergheau, J.M. A new strategy for the numerical modeling of a weld pool. *C. R. Méc.* **2018**, *346*, 999–1017. [CrossRef]
16. Duranton, P.; Devaux, J.; Robin, V.; Gilles, P.; Bergheau, J.M. 3D modelling of multipass welding of a 316L stainless steel pipe. *J. Mater. Proces. Technol.* **2004**, *153–154*, 457–463. [CrossRef]
17. Zain-ul Abdein, M.; Nelias, D.; Jullien, J.F.; Deloison, D. Prediction of laser beam welding-induced distortions and residual stresses by numerical simulation for aeronautic application. *J. Mater. Process. Technol.* **2009**, *209*, 2907–2917. [CrossRef]
18. Feulvarch, E.; Robin, V.; Bergheau, J.M. Thermometallurgical and mechanical modelling of welding—Application to multipass dissimilar metal girth welds. *Sci. Technol. Weld. Join.* **2011**, *16*, 221–231. [CrossRef]
19. Anca, A.; Cardona, A.; Risso, J.; Fachinotti, V.D. Finite element modeling of welding processes. *Appl. Math. Model.* **2011**, *35*, 688–707. [CrossRef]
20. Portelette, L.; Roux, J.C.; Robin, V.; Feulvarch, E. A Gaussian surrogate model for residual stresses induced by orbital multi-pass TIG welding. *Comput. Struct.* **2017**, *183*, 27–37. [CrossRef]
21. Tsirkas, S.; Papanikos, P.; Pericleous, K.; Strusevich, N.; Boitout, F.; Bergheau, J.M. Evaluation of distortions in laser welded shipbuilding parts using local-global finite element approach. *Sci. Technol. Weld. Join.* **2003**, *8*, 79–88. [CrossRef]
22. Lacki, P.; Adamus, K. Numerical simulation of the electron beam welding process. *Comput. Struct.* **2011**, *89*, 977–985. [CrossRef]
23. Piekarska, W.; Kubiak, M. Modeling of thermal phenomena in single laser beam and laser-arc hybrid welding processes using projection method. *Appl. Math. Model.* **2013**, *37*, 2051–2062. [CrossRef]
24. Caprace, J.D.; Fu, G.; Carrara, J.F.; Remes, H.; Shin, S.B. A benchmark study of uncertainness in welding simulation. *Mar. Struct.* **2017**, *56*, 69–84. [CrossRef]

25. Saadlaoui, Y.; Delache, A.; Feulvarch, E.; Leblond, J.B.; Bergheau, J.M. New strategy of solid/fuid coupling during numerical simulation of welding process. *J. Fluids Struct.* **2020**, submitted for publication.
26. Ramaswamy, B. Numerical simulation of unsteady viscous free-surface flow. *J. Comput. Phys.* **1990**, *87*, 495. [CrossRef]
27. Smurov, I.; Doubenskaia, M.; Grigoriev, S. Optical Monitoring in Laser Cladding of Ti6Al4V. *J. Therm. Spray Technol.* **2012**, *21*, 1357–1362. [CrossRef]
28. Doubenskaia, M.; Smurov, I.; Grigoriev, S. Optical Monitoring in Elaboration of Metal Matrix Composites by Direct Metal Deposition. *Phys. Procedia* **2012**, *39*, 767–775. [CrossRef]
29. Kim, C.H.; Ahn, D.C. Coaxial monitoring of keyhole during Yb:YAG laser welding. *Opt. Laser Technol.* **2012**, *44*, 1874–1880. [CrossRef]
30. You, Y.; Gao, X.D.; Katayama, S. Review of laser welding monitoring. *Sci. Technol. Weld. Join.* **2014**, *19*, 181–201. [CrossRef]
31. Doubenskaia, M.A.; Zhirnov, I.V.; Teleshevskiy, V.I.; Bertrand, P.; Smurov, I.Y. Determination of True Temperature in Selective Laser Melting of Metal Powder Using Infrared Camera. *MSF* **2015**, *834*, 93–102. [CrossRef]
32. Lei, T.; Rong, Y.; Xu, J.; Huang, Y. Experiment study and regression analysis of molten pool in laser welding. *Opt. Laser Technol.* **2018**, *108*, 534–541. [CrossRef]
33. Clijsters, S.; Craeghs, T.; Buls, S.; Kempen, K.; Kruth, J.-P. In situ quality control of the selective laser melting process using a high-speed, real-time melt pool monitoring system. *Int. J. Adv. Manuf. Technol.* **2014**, *75*, 1089–1101. [CrossRef]
34. Doubenskaia, M.; Kotoban, D.; Zhirnov, I. Study of oxygen effect on the melting pool temperature during selective laser melting. *Mech. Ind.* **2016**, *17*, 707. [CrossRef]
35. Hooper, P.A. Melt pool temperature and cooling rates in laser powder bed fusion. *Addit. Manuf.* **2018**, *22*, 548–559. [CrossRef]
36. Gao, J.Q.; Qin, G.L.; Yang, J.L.; He, J.G.; Zhang, T.; Wu, C.S. Image processing of weld pool and keyhole in Nd:YAG laser welding of stainless steel based on visual sensing. *Trans. Nonferrous Met. Soc. China* **2011**, *21*, 423–428. [CrossRef]
37. Li, S.; Chen, G.; Zhou, C. Effects of welding parameters on weld geometry during high-power laser welding of thick plate. *Int. J. Adv. Manuf. Technol.* **2015**, *79*, 177–182. [CrossRef]
38. Gao, X.; You, D.; Katayama, S. Infrared image recognition for seam tracking monitoring during fiber laser welding. *Mechatronics* **2012**, *22*, 370–380. [CrossRef]
39. Gao, X.; Zhang, Y. Monitoring of welding status by molten pool morphology during high-power disk laser welding. *Opt. Int. J. Light Electron Opt.* **2015**, *126*, 1797–1802. [CrossRef]
40. Zhang, Y.; Zhang, C.; Tan, L.; Li, S. Coaxial monitoring of the fibre laser lap welding of Zn-coated steel sheets using an auxiliary illuminant. *Opt. Laser Technol.* **2013**, *50*, 167–175. [CrossRef]
41. Li, L.; Peng, G.; Wang, J.; Gong, J.; Meng, S. Numerical and experimental study on keyhole and melt flow dynamics during laser welding of aluminium alloys under subatmospheric pressures. *Int. J. Heat Mass Transf.* **2019**, *133*, 812–826. [CrossRef]
42. Yin, J.; Wang, D.; Yang, L.; Wei, H.; Dong, P.; Ke, L.; Wang, G.; Zhu, H.; Zeng, X. Correlation between forming quality and spatter dynamics in laser powder bed fusion. *Addit. Manuf.* **2020**, *31*, 100958. [CrossRef]
43. Saadlaoui, Y.; Feulvarch, E.; Delache, A.; Leblond, J.B. Apragmatic strategy for the numerical simulation of the fluid flow during welding processes. In Proceedings of the CILAMCE 2018 Congress—XXXIX Ibero-Latin American Congress on Computational Methods in Engineering, Paris/Compiegne, France, 11–14 November 2018.
44. Körner, C.; Attar, E.; Heinl, P. Mesoscopic simulation of selective beam melting processes. *J. Mater. Process. Technol.* **2011**, *211*, 978–987. [CrossRef]
45. Gan, Z.; Yu, G.; He, X.; Li, S. Numerical simulation of thermal behavior and multicomponent mass transfer in direct laser deposition of Co-base alloy on steel. *Int. J. Heat Mass Transf.* **2017**, *104*, 28–38. [CrossRef]
46. Chen, Q.; Guillemot, G.; Gandin, C.A.; Bellet, M. Three-dimensional finite element thermomechanical modeling of additive manufacturing by selective laser melting for ceramic materials. *Addit. Manuf.* **2017**, *16*, 124–137. [CrossRef]
47. Cook, P.; Murphy, A. Simulation of melt pool behaviour during additive manufacturing: Underlying physics and progress. *Addit. Manuf.* **2020**, *31*, 100909. [CrossRef]

48. Smurov, I.; Doubenskaia, M. Temperature Monitoring by Optical Methods in Laser Processing. In *Laser-Assisted Fabrication of Materials*; Majumdar, J.D., Manna, I., Eds.; Series Title: Springer Series in Materials Science; Springer: Berlin/Heidelberg, Germany, 2013; Volume 161, pp. 375–422.
49. Saadlaoui, Y.; Feulvarch, E.; Delache, A.; Leblond, J.B.; Bergheau, J.M. Simulation Numérique des Procédés Thermomécaniques dans une Approche Couplant les Écoulements du Fluide avec les Déformations du Solide: Application au Sudage Laser et à la Fusion d'un lit de Poudre. Ph.D. Thesis, Université de Lyon, Saint Etienne, Frane, 2019.

crystals

MDPI

Article

A Low-Cost Electrochemical Metal 3D Printer Based on a Microfluidic System for Printing Mesoscale Objects

Pengpeng Liu, Yawen Guo, Yihong Wu, Junyan Chen and Yabin Yang *

School of Materials Science and Engineering, Sun Yat-sen University, Guangzhou 510275, China; liupp27@mail2.sysu.edu.cn (P.L.); guoyw26@mail2.sysu.edu.cn (Y.G.); wuyh55@mail2.sysu.edu.cn (Y.W.); chenjy96@mail2.sysu.edu.cn (J.C.)

* Correspondence: yangyabin@mail.sysu.edu.cn

Received: 2 March 2020; Accepted: 23 March 2020; Published: 28 March 2020

Abstract: For the additive manufacturing (AM) of metal objects, the powder-based fusion (PBF) method is routinely utilized to fabricate macroscale parts. On the other hand, electrochemical additive manufacturing (ECAM), in which metallic structures are deposited through the electrochemical reduction of metal ions, is a promising technique for producing micro- and nanoscale objects. However, a gap exists in terms of fabricating mesoscale objects within the current AM techniques. The PBF method is limited by fabrication precision due to pronounced residual stresses, and most current ECAM systems are difficult to scale up to print mesoscale objects. In the present paper, the novel design of a low-cost ECAM 3D printer based on a microfluidic system is proposed for fabricating mesoscale metal parts. The meniscus-guided electrodeposition approach is utilized, in which a meniscus is formed between the print head and substrate, and electrodeposition is confined within the meniscus. A 3D object is fabricated by the meniscus moving with the print head according to the programmed pattern and the material subsequently being deposited at the designated locations. The key to the proposed design is to maintain a mesoscale meniscus, which normally cannot be sustained by the electrolyte surface tension with a print nozzle having a mesoscale diameter. Therefore, a microfluidic system, called the fountain pen feed system, constituting a semi-open main channel and comb structure, was designed to maintain a mesoscale meniscus throughout the printing process. Two materials, copper and nickel, with various geometric shapes were attempted to print by the proposed ECAM system, and, during the printing process, both fluid leaking and meniscus breaking were completely prevented. Free standing tilted copper pillars with controlled angles were printed to show the ability of the proposed design in fabricating 3D structures. A copper circuit was also printed on a non-conductive substrate to demonstrate a possible application of the proposed ECAM system in the fabrication of functional electronics.

Keywords: electrochemical additive manufacturing; fountain pen feed system; metal 3D printer

1. Introduction

Additive manufacturing (AM) (3D printing), the method of creating geometrically complex 3D parts in a layer-by-layer manner, enables the fabrication of complex parts with high topological freedoms within a single manufacturing step. For metals, macroscale parts are usually fabricated using the powder-based fusion (PBF) method, which involves a high-energy external heat source such as a laser or electron beam to selectively melt metallic powders and fuse them together according to a computationally programmed pattern [1,2]. For microscale and nanoscale structures, electrochemical additive manufacturing (ECAM), in which metal ions are deposited as metal atoms at given locations by an electrochemical process, has been suggested as a promising approach to achieve excellent mechanical and electrical properties with extremely high resolution [3,4]. However, there is a gap in terms of cost-effective AM approaches to print mesoscale structures in the sub-millimeter range.

The resolution of the PBF method is limited by the powder diameter, which is in the range of 20–100 µm [5]. Additionally, high residual stress induced by thermal deformation [6,7] is pronounced in PBF, which considerably impairs the printing precision. For ECAM, two approaches are mainly utilized:

1. Localized electrodepotion (LED), for which a conductive substrate and an ultrasharp microelectrode are immersed into a bath of metal salt solution (electrolyte). Localized deposition of metals occurs through the electrochemical reduction when the microelectrode is very close to the substrate and an external electric field is applied between the microelectrode and substrate [8–10].
2. Meniscus-guided electrodeposition (MGED) [11], for which a micropipette that contains the metal salt solution is utilized. When the micropipette approaches the conductive substrate, a liquid meniscus is established. The electrodeposition is then confined to the small region of the meniscus when applying appropriate potential difference [12].

As in MGED, the printing process is operated in air, and the printed structures will not be contaminated or destroyed by submersion [11]. Seol et al. [4] demonstrated geometrically complex microarchitectures can be printed using the MGED approach by modulating the applied potential in two steps with different amplitudes and durations. By utilizing a pulsed potential, Behroozfar et al. [3] successfully printed nanotwinned-metals, which exhibit superior mechanical and electrical properties. Kim et al. [13,14] developed an AFM-based nanofountain probe with the form of volcano tip, which is capable of delivering a meniscus at the sub-100 nm scale. This design is then successfully applied for multiple applications [15–18]. Besides, Falola et al. [19] introduced a wide range of elements that can be electrodeposited from aqueous electrolytes. Reiser et al. [20] printed bi-metal (Cu-Ag) pillars by controlling the electrohydrodynamic ejection of metal ions dissolved from the Cu or Ag anode. Chen et al. [21] developed a two-syringe system, one for copper and one for nickel, to fabricate thermally responsive Cu-Ni strips. Ambrosi et al. [22] demonstrated how the materials (metal copper and polyaniline) to be deposited within a single syringe can be precisely controlled by tuning the applied potential.

Although these methods demonstrate they are valid alternatives to conventional ink-based or laser-based AM technologies, the products are mainly limited to micro- or nanoscale applications. Furthermore, expensive piezo-driven positioning stages are usually required to maintain a stable meniscus [22]. Therefore, a more economically efficient metal printing technology to fabricate large-scale products is highly needed. The key problem is how to maintain a stable mesoscale meniscus, which cannot be obtained merely by the surface tension of the liquid. Wang et al. [23] modified a syringe extruder to print 2D patterns of Cu_2O. A controlled volume of liquid electrolyte was ejected to form a meniscus between the print head and substrate. Kim et al. [24] introduced a suction nozzle at the print head to maintain a stable meniscus. Chen et al. [25] proposed an ECAM system based on low-cost stepper motor stages instead of expensive piezo-based stages. A nozzle with a diameter of 400 µm was employed to increase the size of the meniscus considerably and thus improve the deposition rates. In order to maintain a stable meniscus with such a large diameter, they proposed inserting a porous sponge at the tip of the nozzle to adjust the electrolyte flow rate. They further improved their approach by using porous electrospun nanofiber mats at the tip instead of a sponge to print Cu-Ni strips [21]. However, in their approach, the flow rate of the electrolyte solution was intrinsically controlled by the permeability of the porous material, which may vary during the printing process and hence cause certain difficulties in maintaining a stable meniscus. Moreover, precise control over insertion of the porous material at the desired nozzle position was also difficult. Nevertheless, they demonstrated that mesoscale meniscus-confined ECAM is possible.

In the present paper, we propose a novel ECAM system design based on the MGED approach for printing mesoscale metal objects. A microfluidic system referred to as the fountain pen feed system was employed to maintain a stable mesoscale meniscus. With the proposed design, passive control of continuous electrolyte flow at the tip of the print head can be achieved without liquid leaking or breaking during the printing process, and thus the stability of the meniscus can be automatically adjusted by the fountain pen feed system. As there is no need for active control, the proposed fountain pen feed system is convenient to establish. Moreover, the proposed fountain pen feed system opens the door to controlling the profile of the meniscus with various designs of the microfluidic system. A fluid flow simulation performed on the commercial software Fluent was utilized to help to explain the mechanism of the fountain pen feed system. Two materials, copper and nickel, with various geometric shapes were attempted to print to demonstrate the feasibility of the proposed ECAM system. A printing process on a non-conductive substrate was also conducted to provide a possible application with the proposed design. The article concludes with a reiteration of the most salient points of the study.

2. Design

The proposed ECAM system is shown in Figure 1a. Movement of the print head is controlled by the $x-$, $y-$, and $z-$ stages. The design of the print head is schematically illustrated in Figure 1b. A microfluidic system, which is referred to as the fountain pen feed system, is proposed. This fountain pen feed system consists of the main channel part and the comb structure, which is further illustrated in details in Figure 2a. During the printing process, the electrolyte in the reservoir flows through the main channel to the tip by gravity and capillary action. At the same time, with the semi-open design of the main channel (as shown in the A–A cross-section of Figure 1b), air can also flow through the main channel into the reservoir, thereby providing a sufficient back pressure for continuous transport of electrolyte to the tip of the print head. When a positive potential is applied between the working electrode and substrate, the metal ions M^{n+} within the meniscus are deposited on the substrate through the reduction of $M^{n+} + ne^- = M$. The meniscus moves with the print head according to a programmed path, and the materials are subsequently deposited at the designated locations to form the desired structure.

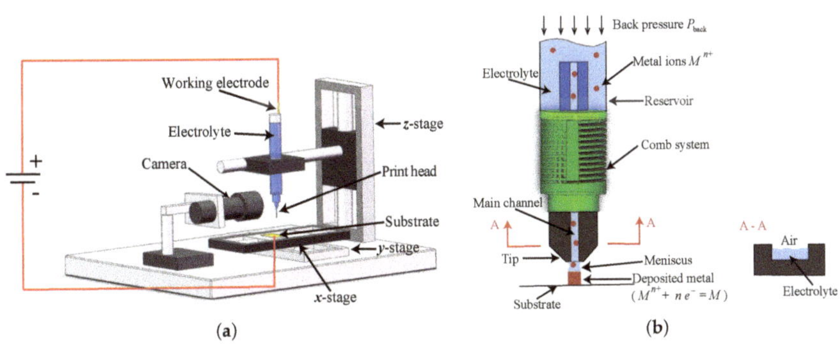

Figure 1. (**a**) Schematic illustration of the proposed ECAM system. (**b**) Detailed illustration of the print head, also referred to as the fountain pen feed system. The cross section A–A at the tip of the main channel is shown in the lower right corner.

As shown in Figure 2a, there is a slit in the comb structure. Therefore, if disturbances occur, such as a variance in temperature or unevenness of the substrate, the excess liquid which passes into the main channel from the reservoir can flow through the slit of the comb structure to the outside branches, and thus liquid leaking can be prevented at the tip of the print head.

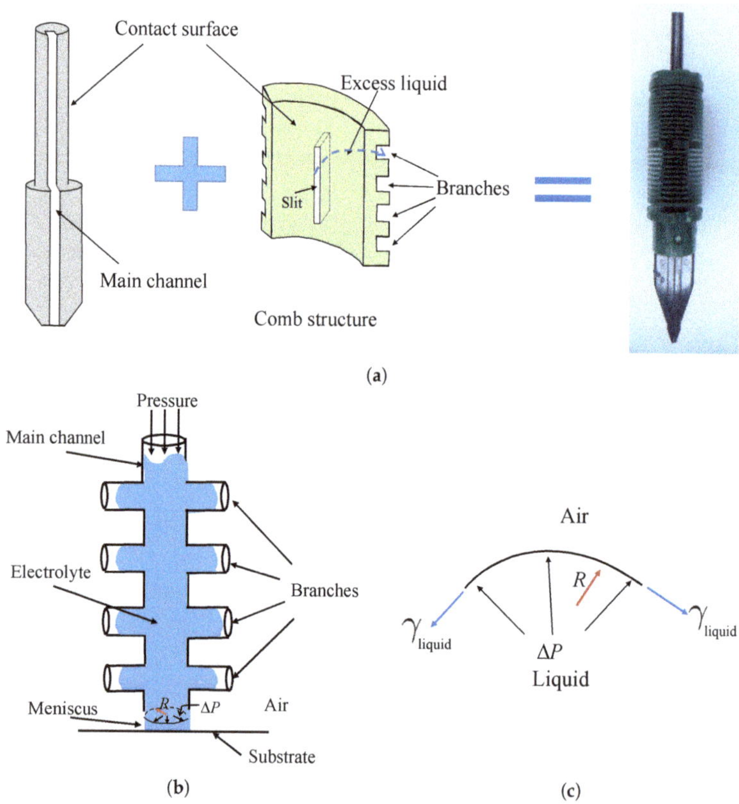

Figure 2. (**a**) The detailed illustration of the main channel and the comb structure. Excess fluid can flow through the slit of the comb structure and to the outside branches. (**b**) Schematic illustration that how the branches can help to main a stable meniscus. (**c**) Force diagram of the meniscus at the interface between liquid and air.

This mechanism is further explained in Figure 2b,c. The branches shown in Figure 2b act as the comb structure to store the excess liquid. If the meniscus shown in Figure 2b is simplified to have a shape of cylinder with the radius of R, which is equal to the radius of the main channel, the key to maintaining a stable meniscus is the pressure difference between the inside and outside of the meniscus, which can be balanced by the surface tension γ_{liquid} of the liquid, as illustrated in Figure 2c. The force equilibrium is expressed as

$$\Delta P = \frac{\gamma_{\text{liquid}}}{R}. \quad (1)$$

If disturbances occur which causes a sudden increase of ΔP, and the surface tension γ_{liquid} is insufficient to balance this increase, liquid leaking would occur at the tip of the print head. On the other hand, if the radius R is increased to enlarge the deposition area, which thus improves the deposition rate, the pressure provided by the surface tension γ_{liquid} would decrease, and hence the stability of the meniscus is impaired. The branches shown in Figure 2b not only store the excess liquid but also prevent the abrupt increase of the pressure difference ΔP at the tip of the print head. The effect of adding the branches was further analyzed by a simulation run in the commercial software Fluent.

As shown in Figure 3a,b, two 2D simulations were performed. There are no branches for the simulation in Figure 3a, and for the simulation in Figure 3b, eight branches are located symmetrically with respect to the main channel. The width of the branch channel is 20 μm and the distance between adjacent branch channels is 30 μm. A constant pressure of $\Delta P = 2000$ Pa was applied at the inlet for both simulations shown in Figure 3a,b. The volume of fluid (VOF) approach in Fluent was used to track the interface of the liquid and air. Water was taken as the liquid in the simulations and the material parameters employed in the simulations are tabulated in Table 1. For the phase field ϕ_v shown in Figure 3a,b, air is represented by $\phi_v = 0$ while water is represented by $\phi_v = 1$. Rectangular elements with the mesh size of 0.004 mm × 0.004 mm are used and the time increment is 10^{-5} s.

Figure 3a,b shows the phase filed at a certain time for the case without and with eight branches, respectively. $\phi_v = 0.8$ is chosen as the interface of water and air. It can be observed that the meniscus for the case without branches tends to flow outward while the meniscus for the case with eight branches is maintained in reasonable shape. The pressure from Point A to Point B (see Figure 3a) for the two cases is compared in Figure 3c. It can be clearly observed that the pressure for the case with eight branches is lower than that without branches, which implies the stability of the meniscus can be improved by adding the branches. It should be noted that the simulation with eight branches shown in Figure 3b is employed to schematically illustrate that the branched structure is capable of maintaining a more stable meniscus. There are 24 branches for the real comb structure shown in Figure 2a. Therefore, the effect of maintaining a stable meniscus by the real comb structure is expected to be much better than that of the simulated structure with eight branches. As detailed in the following sections, several experiments were conducted to demonstrate the processing ability of the proposed design.

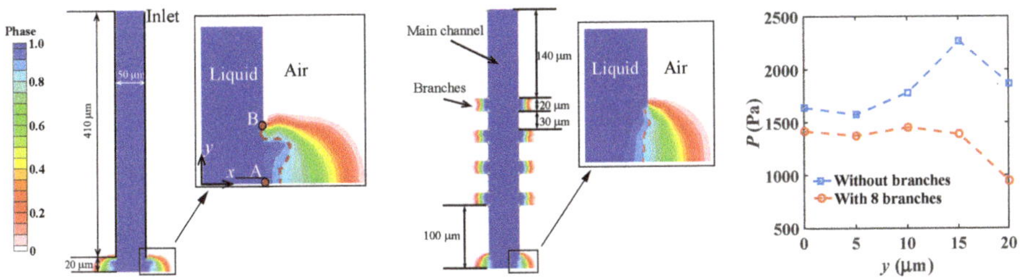

(a) (b) (c)

Figure 3. Simulations on the fluid flow in the main channel: (**a**) without branches; and (**b**) with eight branches. (**c**) The pressure along Point A to Point B shown in (**a**).

Table 1. Physical conditions of water and air [26].

	Density (kg/m^3)	Viscosity (Pa·s)	Surface Tension (N/m)
Water	988.2	0.001	0.072
Air	1.225	1.8×10^{-5}	-

3. Experiments

3.1. Materials

The experiments were mainly conducted for the deposition of copper and the electrolytes were a mixture of 1 M $CuSO_4$ and 0.005 M H_2SO_4 in deionized water with ratio 1:1 by volume. A nickel square was also printed to demonstrate the ability to print various materials. The electrolytes for nickel deposition were 113 gL^{-1} $NiSO_4 \cdot 6H_2O$, 30 gL^{-1} $NiCl_2 \cdot 6H_2O$ and 23 gL^{-1} H_3BO_3 [21].

3.2. Apparatus

Three two-phase stepper motor stages (Zolix, Beijing) were employed to serve as the x-, y-, and z-stages. The stepper motor stages have the screw pith of L = 4 mm and step angle of $\alpha = 1.8^\circ$ with n = 20 subdivisions, resulting in the positioning resolution of $L\alpha/360n = 1$ μm in Cartesian coordinates.

Three materials, namely the polished brass plate which was composed of 67%Cu and 33%Zn with the surface roughness Ra = 0.017 μm, the unpolished copper plate with Ra = 0.19 μm, and the polyethylene terephthalate (PET), were tested to serve as the substrate. The surface roughness Ra, which was evaluated by the arithmetic average of the absolute values of the profile height deviations from the mean line within the evaluation length, was measured using a Mitutoyo SJ-210 surface roughness tester with an evaluation length of 1 mm.

As shown in Figure 1a, a syringe was employed as the reservoir and a copper or nickel wire was inserted into the electrolyte to serve as the anode, while the substrate served as the cathode. The electrical potential was applied between the anode and cathode by a signal generator (Tektronix AFG3011C). The syringe with the print head was held onto the z-stage and the print head was lowered slowly until the tip touched the substrate where a current signal was recorded by a Keithley digital multimeter DMM7510. Then, the print head retracted 30 μm to form a stable meniscus.

A two-electrode system without the reference electrode was used in the present study. Since the active area of the anode (counter electrode) was far larger than that of the cathode (working electrode), the polarization of the counter electrode was expected to be small [27]. The deposition voltage was tuned to make the corresponding current ∼0.3 mA, which was reported as the optimal deposition current for copper in [22]. After a parametric study, the potential of 25 V between the anode and cathode was selected (the corresponding current recorded was ∼0.3 mA). Both the constant (25 V) and pulsed potential (25 V, 1 kHz, duty cycle = 0.3) were tested for copper deposition. For simplicity, a content potential of 25 V was also employed to print the nickel. It should be noted that the potential value selected was relatively larger than that reported in [25], in which a three-electrode system where a copper bar was used as the reference electrode was employed and a maximum potential of 6 V was applied between the reference electrode and the cathode. In the proposed design, the electrolyte needs to pass through the long and narrow main channel to the tip of the print head. As a result, the transport of the metal ions from the reservoir to the tip of the print head is limited due to the high width to height ratio of the main channel [28]. Therefore, the concentration of the metal ions through the long main channel is expected to be lower than that in the reservoir, which would increase the resistance of the whole circuit.

A commercially obtained fountain pen, as shown in Figure 4, was employed as the print head. The tip of the main channel was machined by a laser engraving machine (Branded Hans Laser) to control the width in the range of 50–300 μm. The width of the slit in the comb structure was ∼400 μm. The material used for the main channel and the comb structure was acrylonitrile butadiene styrene plastic (ABS).

Scanning electron microscopy (SEM) images and energy-dispersive X-ray spectroscopy (EDX) element mapping were obtained at high vacuum with operating voltage of 20 kV. Elastic modulus of the printed structure was measured by the nanoindentation test on an Agilent Nano Indenter G200, and microhardness tests were performed on a Fischerscope HM200.

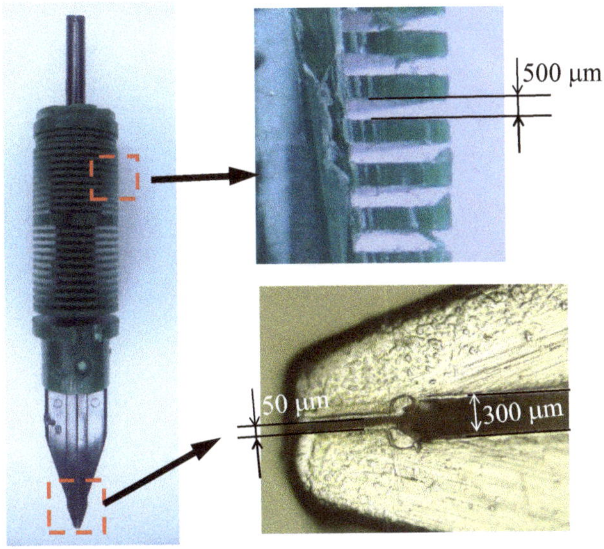

Figure 4. The dimensions of the print head.

4. Results and Discussions

4.1. Printing Speed

To determine the maximum appropriate printing speed for the proposed ECAM system that can be achieved with a relatively high resolution and uniform morphology, lines printed at various printing speeds on the brass substrate were investigated. For simplification and convenience regarding the influence of frequency and duty cycle of the pulsed potential, a constant potential was simply applied. An optical microscope was employed to measure the printed line width. Figure 5a shows the printed copper line width as a function of printing speed. It can be observed that the printed line width decreases in an approximately linear fashion as speed increases, which indicates that a higher speed leads to a higher resolution of the printed line. However, a printing speed that is too high results in a non-uniform printed line width along the printing direction and serrated boundaries. For instance, as shown in Figure 5a, the line printed at 500 µm/s is much more uniform than that printed at 1000 µm/s. Further increasing the printing speed resulted in non-continuous printed lines, as shown in Figure 5b for the speed of 3000 µm/s. On the other hand, a printing speed that is too low not only impairs the deposition rate but also allows sufficient time for solvent (water) evaporation. As a result, the solute material (for instance, $CuSO_4$ for copper deposition) readily precipitates on the substrate, such as for the line printed at 100 µm/s. Consequently, in the present study, 500 µm/s was chosen as the optimal speed.

For the speed of 500 µm/s, taking the diameter of the meniscus having the same dimension, i.e., 500 µm, and the current to be $\sim$ 0.3 mA, then the corresponding current density was 0.15A/cm^2, which was comparable to the values reported in [3,21,29] (0.02–0.64 A/cm^2).

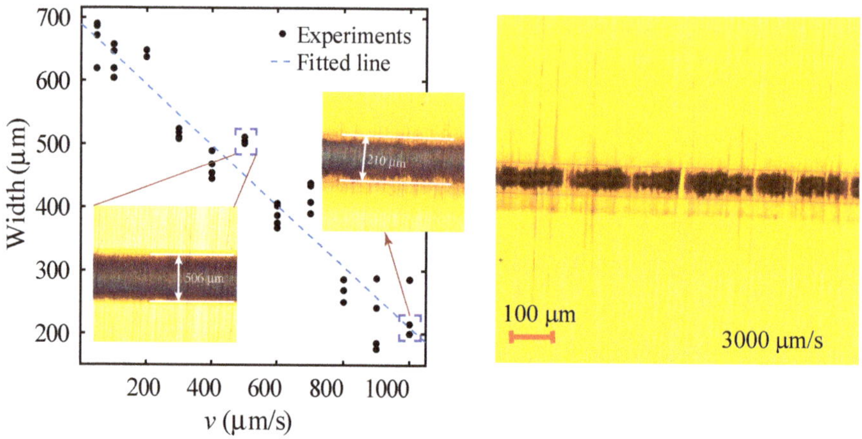

(a) (b)

Figure 5. (**a**) Printed line width as a function of printing speed. (**b**) An optical microscope image of the printed line with the speed of 3000 µm/s.

4.2. Printing a Copper Square

A copper square with the length of 2 mm was printed on the brass substrate. Two printing strategies, as shown in Figure 6a,b, with the constant and pulsed potential were investigated. The distance between two adjacent printing lines was 100 µm. For Printing Strategy 1 (see Figure 6a), the print head moves along the x-axis for all the layers; Ra_x in Figure 6a is the surface roughness measured along the x-direction, which is parallel to the printing direction, while Ra_y is the surface roughness measured along the y-direction, which is perpendicular to the printing direction. For Printing Strategy 2 (see Figure 6b), the printing direction rotates 90° after printing each layer, and the surface roughness Ra is measured for both the x- and y-directions. The Ra values shown in Figure 6c were averaged by measuring three samples, and, for each sample, at least five different locations were analyzed.

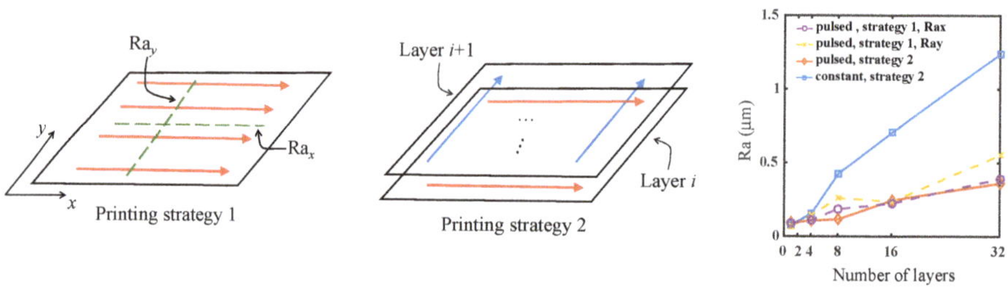

(a) (b) (c)

Figure 6. The average surface roughness, Ra, was investigated for two printing strategies. (**a**) Printing Strategy 1, where the print head moves along the x-axis for all layers. (**b**) Printing Strategy 2, where the printing direction rotates 90° after printing each layer. (**c**) Ra as a function of the number of deposition layers and printing strategies under pulsed and constant potential.

Figure 6c shows that Ra increases with the increasing number of deposition layers. For Printing Strategy 1, Ra_x is slightly smaller than Ra_y, which suggests that the surface is smoother along the direction parallel to the printing direction than that along the perpendicular direction. In contrast, the square obtained by Printing Strategy 2 with a pulsed potential has the smallest Ra, which indicates that the surface roughness can be improved by applying the pulsed potential with varying the printing directions during the process. The advantage of Printing Strategy 2 compared to Printing Strategy 1 is that the upper-layer material can be filled into the gap and hole of the lower layer during the printing process, therefore some uneven areas can be repaired and the surface flatness is improved. The surface roughness shown in Figure 6c was measured using a Mitutoyo SJ-210 surface roughness tester with an evaluation length of 1 mm.

Figure 7a shows the printed square by depositing two layers on the brass substrate with Printing Strategy 2 and the pulsed potential. The yellow part at the boundary is believed that some electrolytes were accumulated at the boundary during printing and the zinc element in the brass substrate was dissolved in the electrolytes and then deposited on the substrate. This was confirmed by using an unpolished copper substrate and, repeating the same process for printing the square, the yellow part was no longer observed, as shown in Figure 7b. The proposed ECAM system also worked well on both the polished and unpolished substrates, and the stability of the meniscus was not affected. Further measurements showed that the square printed on the polished substrate resulted in a smoother printed surface (Ra = 0.036 μm) than that on the unpolished substrate (Ra = 0.233 μm). This is mainly because the actual surface area of the rough surface is larger than that of the smooth surface, therefore the current density on the rough surface is relatively small. If the local current density were not sufficient for the deposition of metal ions, there would be no metals deposited on that region. Hence, due to the worse uniformity of current density on the rough surface, the corresponding roughness of the printed structures is higher.

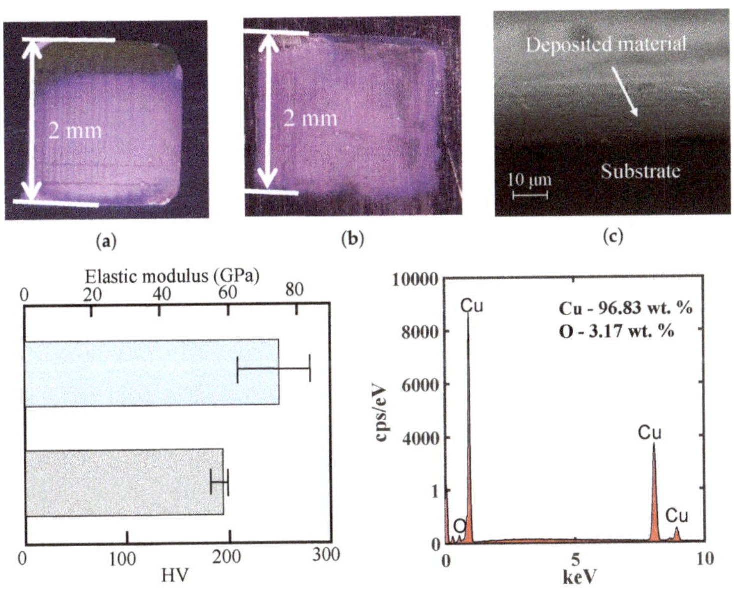

Figure 7. (**a**) A copper square printed on the polished brass substrate for two layers. (**b**) A copper square printed on the unpolished copper substrate for two layers. (**c**) SEM image of the side view of a copper square printed on the brass substrate for 32 layers. (**d**) Elastic modulus and Vickers hardness for the printed copper square. (**e**) EDX of the printed copper square.

The side view for the square printed on the brass substrate for 32 layers is shown in Figure 7c, illustrating the dense morphology and height of ~12 μm. As the printing speed was 500 μm/s with 22 passes (each pass was 2 mm) for printing each layer, the total time for printing every layer was 88 s. Hence, Figure 7c suggests the thickness for each pass is ~375 nm. This implies that the printing process has an exceptionally high z-height resolution, making it potentially useful for fabricating functional electronics such as thin-film sensors.

The elastic modulus *E* and Vickers hardness HV of the printed copper square on the polished brass substrate were measured by the nanoindentation and microhardness tests, respectively. Six points were tested and the results are shown in Figure 7d. It can be observed that the average elastic modulus of the printed copper is ~75 GPa, while the average Vickers hardness is ~200 HV. As a reference, Chen et al. [25] reported that the hardness of their printed copper structures ranged from 184 to 228 HV and the hardness of copper used in circuit bonding wires can range from 50 to 176 HV [30]. The processes for obtaining the elastic modulus and Vickers hardness are explained in Appendix A. The formation of pure elemental copper for the printed square on the polished brass substrate was confirmed by EDX, as shown in Figure 7e. The deposited copper of the test point can be as high as 96.83 wt% with 3.17 wt% of oxygen. Further SEM study showed the grain size of the printed square was as small as ~1 µm.

4.3. Printing Various Copper Structures

Three letters, "S", "Y", and "U", were separately printed from corresponding computer-aided design (CAD) models with pulsed potential and the image of a combination of the printed letters featuring the pattern "SYSU" is shown in Figure 8a. As the letters were printed separately, to make a uniform background, the substrate is removed and only the letters are shown in Figure 8a. A typical SEM image of the letter "U" is shown in Figure 8b, illustrating the dense copper morphology without noticeable dendrites. Figure 8c is a close-up of the white boxed area in Figure 8b and clearly shows that the printed letter has a smooth surface. Multiple passes were applied for printing each letter and the printing time was ~45 min. SEM images of the printed letter "Y" (Figure 7d) indicates that the printed copper structures are polycrystalline with the grain size as small as ~1 µm. The letter "Y" shown in the boxed area in Figure 7d is an unedited image without removing the substrate.

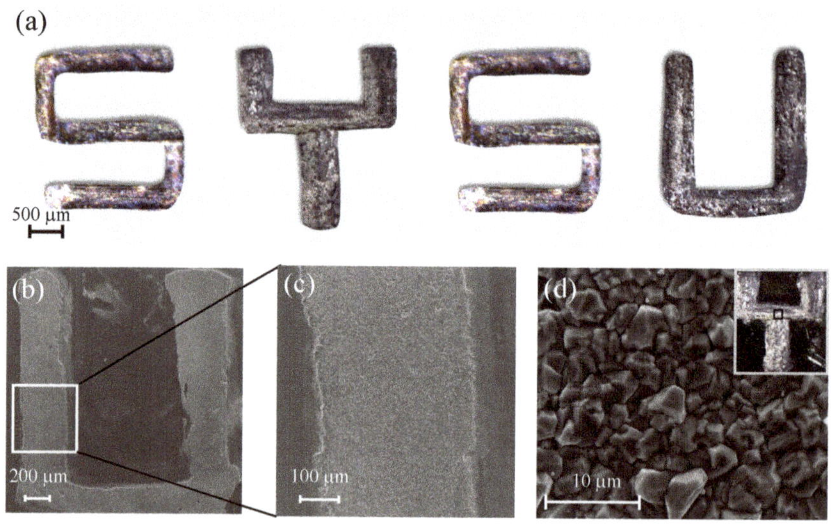

Figure 8. (**a**) Images of printed structures featuring the letters "SYSU." Only the letters without the substrate are shown to make an uniform background. (**b**) A typical SEM image of the letter "U"; the zoomed-in view of the white boxed area is shown in (**c**). (**d**) SEM image of the white boxed area in letter "Y".

The letter "U" printed under pulsed and constant potential are also compared in Figure 9a,b. The SEM images show that the pulsed potential results in denser morphologies with sharp boundaries (see Figure 9a), whereas the structures printed under the constant potential have a porous morphology (see

Figure 7b). This is mainly because the pulsed potential can provide a higher current density, which favors the initiation of grain nuclei and leads to finer-grained deposition [31]. Again, The two letters "U" shown in the boxed area of Figure 9a,b are unedited images without removing the substrate.

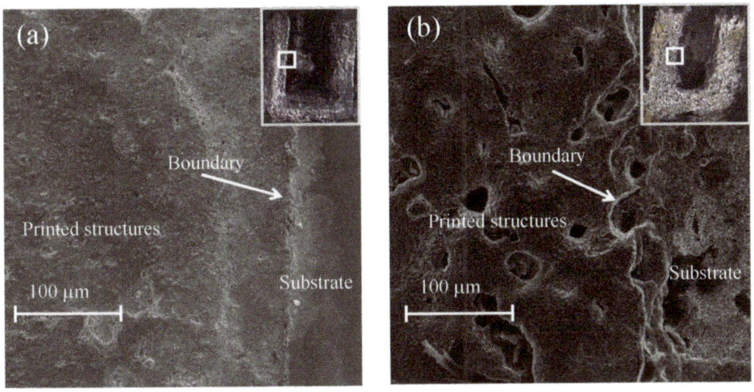

Figure 9. SEM images of the letter "U" printed under (**a**) pulsed and (**b**) constant potential.

A few attempts were also made to print free standing copper pillars, which are shown in Figure 10. Figure 10a,b shows the printed copper pillars with slopes of $\tan\alpha = 1$ and 2, respectively, where α is the angle with respect to the x-axis. Figure 10c is a printed vertical pillar followed by a pillar with the slope $\tan\alpha = 3$, and Figure 10d is a mixed structure with the overhanging pillars. The red arrows indicate the movement paths of the print head, and the numbers in Figure 10d represent the printing path sequence. The printed pillars with slopes $\tan\alpha$ in Figure 10a–c were achieved by alternately moving the print head a distance of $d = 3$ μm along the x-axis and $d \cdot \tan\alpha$ along the z-axis. The movement paths in Figure 10d were also realized by controlling the movement distances of the x- and z-stages via a computer program. The speed of the print head was controlled to a low value (3 μm/s) for both the x- and z-stages. The diameter of the printed pillars in Figure 10 is ∼50 μm, and dendritic morphologies can be observed in all structures. As explained in [25], when printing structures such as those in Figure 8, movement of the print head within the x–y plane results in mechanical removal of the dendrites. However, for the structures shown in Figure 10, without the aid of mechanical removal, dendrites are easily formed due to preferential deposition at the center of the meniscus, which could be improved by applying an appropriate type of potential [4,25]. Nevertheless, Figure 10 demonstrates the ability of the proposed ECAM system to print structures with small overhanging angles.

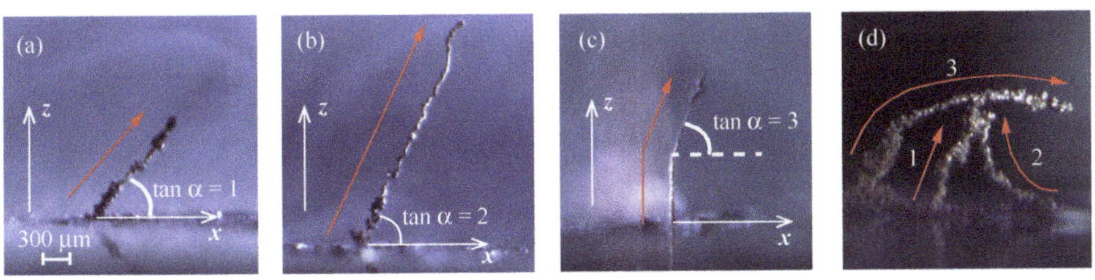

Figure 10. Images of 3D-printed Cu structures with different shapes. Red arrows indicate the movement paths of the print head. (**a**) A tilted pillar with slope $\tan\alpha = 1$, where α is the angle with respect to the x-axis. (**b**) A tilted pillar with slope $\tan\alpha = 2$. (**c**) A vertical pillar followed by a tilted pillar with slope $\tan\alpha = 3$. (**d**) Mixed structure with overhanging features. The numbers indicate the sequence of the printing path.

4.4. Printing a Copper Circuit on the PET Substrate

A circuit printed on the PET substrate was attempted to present a potential application with the proposed ECAM system. As the PET substrate was non-conductive, a thin copper film was first deposited on the PET substrate as the seed layer through the vacuum vapor plating technology. The thickness of the seed layer was 150 nm. This thickness ensured good chemical and mechanical properties of the deposited copper [32] and good adhesion between the substrate and the seed layer [24]. The proposed printing process was then performed on the seed layer to fabricate the designed patterns. Finally, a chemical etching process was employed to remove the seed layer and clean the substrate. The etching solution was composed of 0.4 M hydrochloric acid and 0.07 M $FeCl_3$ and the etching process lasted 10 s. Before depositing the seed layer, the substrate was cleaned by rinsing with ethanol and ultrapure water and subsequent 10 min plasma. As shown in Figure 11, a "pulse signal" pattern was printed on the PET substrate. Multiple passes were performed for printing the "pulse signal" pattern and the printing time was ~1 h. This example is shown as a proof-of-concept. Further investigations are needed to improve the quality of the printed circuit by optimizing the processing parameters and testing the corresponding electrical and mechanical properties.

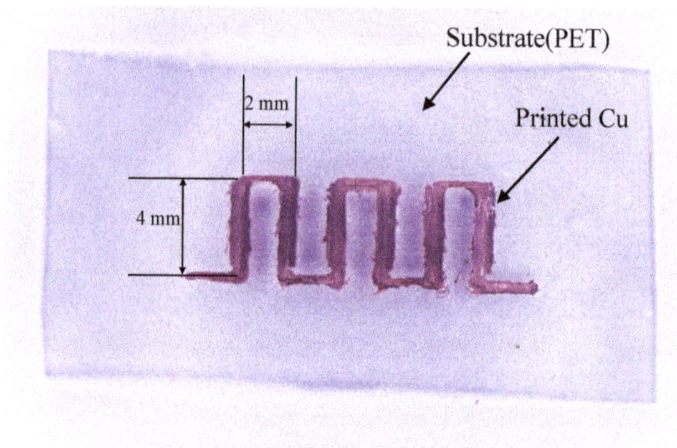

Figure 11. A "pulse signal" pattern printed on the PET substrate.

4.5. Printing a Nickel Square

A nickel square was finally printed with the same configurations as printing the copper. The square was printed using Printing Strategy 2 (see Figure 6b) for 16 layers. The SEM image of the printed nickel square is shown in Figure 12a. The formation of the nickel was further confirmed by testing a small region of the printed square using the EDX element mapping, as shown in Figure 12b. The printed nickel was 86.47 wt%. The elements of copper and zinc tabulated in Figure 12b were believed to come from the brass substrate. The surface roughness Ra of the printed nickel square was 0.141 μm. Figure 12 demonstrates various materials can be printed by the proposed ECAM system. Future studies may incorporate the electrolytes for printing the nickel and copper and control the depositing sequence of different elements by manipulating the applied potential.

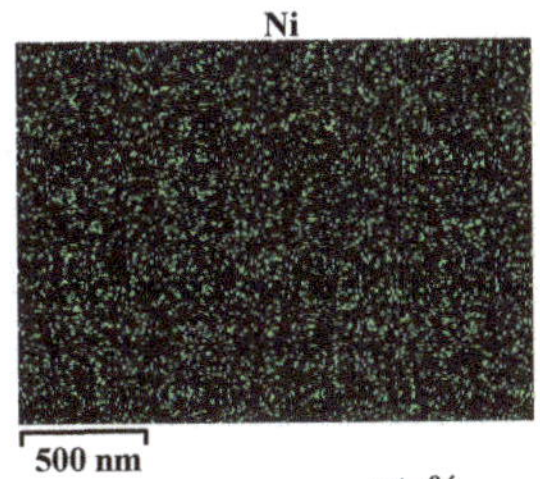

Elements	wt. %
Ni	86.47
Cu	8.4
Zn	4.32
O	0.81

(a) (b)

Figure 12. (**a**) The printed nickel square. (**b**) EDX element mapping for a small region of the printed nickel square and the corresponding wt.% of the detected elements.

5. Conclusions

This work presents the novel design of a low-cost ECAM system for printing mesoscale metal parts. With the proposed microfluidic system, i.e., the fountain pen feed system, which is composed of a semi-open main channel and comb structure, a stable mesoscale meniscus can be obtained during the printing process without liquid leaking or meniscus breaking. Two materials, copper and nickel, were attempted to print. It is believed that some other metals that can be electrodeposited from aqueous electrolytes, such as Ag and Ti, can be printed through the same ECAM system. The investigation of the surface roughness of the printed structures shows that the surface smoothness can be controlled by manipulating the applied potential and printing strategy. Various structures were also printed and an example of a copper circuit printed on the non-conductive PET substrate was achieved to provide a possible application using the proposed ECAM system. The presented work demonstrates the possibility to control the meniscus and electrolyte flow by designing an appropriate microfluidic system and provides a foundation for future work on designing more sophisticated microfluidic systems to achieve higher printing precisions and satisfactory material properties for the fabrication of functional electronics.

Author Contributions: Conceptualization, Y.Y.; validation, P.L., Y.G., Y.W., and J.C.; formal analysis, P.L. and Y.G.; investigation, P.L. and Y.G.; writing–original draft preparation, P.L. and Y.Y.; writing–review and editing, P.L., Y.G., Y.W., J.C., and Y.Y.; and supervision, Y.Y. All authors have read and agreed to the published version of the manuscript.

Funding: This research received no external fund.

Conflicts of Interest: The authors declare no conflict of interest.

Appendix A.

Appendix A.1. Nanoindentation Test

A load of 0.78 mN was applied on the sample for the nanoindentation test with a Berkovich indenter and a typical loading and unloading curve is shown in Figure A1. The unloading curve was fitted by

$$P_u = a(h - h_f)^m, \tag{A1}$$

where a, h and h_f are the fitting parameters, which were fitted using MATLAB in the present study. Then, the contact stiffness was established as

$$S = \frac{dP_u}{dh}|_h = h_{\max} = ma(h_{\max} - h_f)^{m-1}, \tag{A2}$$

where $h_{\max}$ is the maximum indentation depth.

The elastic modulus and hardness were calculated by the Oliver–Pharr method and were expressed as

$$\frac{1}{E_r} = \frac{1 - v^2}{E} + \frac{1 - v_i^2}{E_i}, \tag{A3}$$

where E_i and v_i are the elastic modulus (1140 GPa) and Poisson's ratio (0.07), respectively, of the indenter, and E and v are the elastic modulus and Poisson's ratio, respectively, of the test material. The modulus E_r was expressed as

$$E_r = \frac{\sqrt{\pi}}{2} \cdot \frac{S}{\sqrt{A}}, \tag{A4}$$

where S is the contact stiffness and A is the projected contact area. The projected contact area A was calculated by

$$A = C_1 h_c^2 + C_2 h_c + C_3 h_c^{1/2}, \tag{A5}$$

where C_1 = 23.97, C_2 = 391.7, and C_3 = 2018.2, which were calibrated by the test machine. The contact depth h_c was estimated by

$$h_c = h_{\max} - \varepsilon \frac{P_u}{S}, \tag{A6}$$

where ε is a constant, which is related to the geometry of the indenter. For a Berkovich indenter, $\varepsilon = 0.75$.

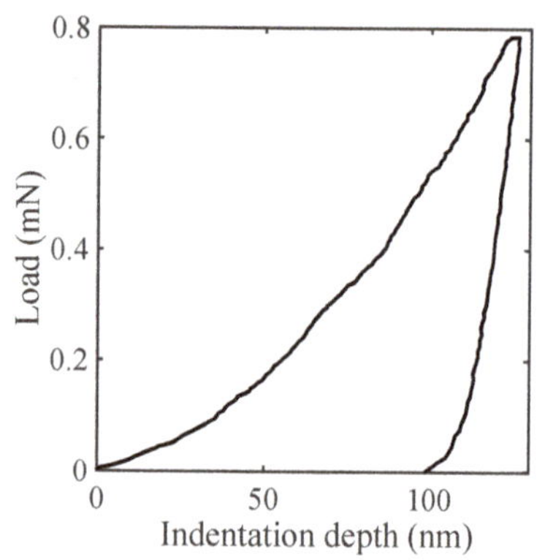

Figure A1. The loading and unloading curve from the nanoindentation test.

For all the material points tested, the ratio of final indentation depth h_f to the maximum indentation depth $h_{\max}$ ($h_f / h_{\max}$) was ~0.83, which indicated that slight pile-up may have occurred under nanoindentation. As a result, the measured elastic modulus and hardness may be slightly overestimated. As reported by Moharrami and Bull [33], for the indentation depth obtained in the present study, the maximum overestimation for the elastic modulus was 8%.

Appendix A.2. Vickers Hardness Test

A maximum load of F = 1000 mN was applied for the Vickers hardness test with the maximum indentation depth of ∼4.6 μm. The loading and unloading time were both 20 s and the dwell time was 10 s. For the Vickers hardness test, a diamond indenter with the tip angle of 136° was employed and the averaged diagonal length d (μm) of the contact area was measured to calculate the Vickers hardness HV by

$$\mathrm{HV} = \frac{0.1891F}{d^2} \times 10^3. \quad \text{(A7)}$$

References

1. Yap, C.Y.; Chua, C.K.; Dong, Z.L.; Liu, Z.H.; Zhang, D.Q.; Loh, L.E.; Sing, S.L. Review of selective laser melting: Materials and applications. *Appl. Phys. Rev.* **2015**, *2*, 041101. [CrossRef]
2. Gong, X.; Anderson, T.; Chou, K. Review on powder-based electron beam additive manufacturing technology. In Proceedings of the ASME/ISCIE 2012 International Symposium on Flexible Automation, St. Louis, MO, USA, 18–20 June 2012; pp. 507–515.
3. Behroozfar, A.; Daryadel, S.; Morsali, S.R.; Moreno, S.; Baniasadi, M.; Bernal, R.A.; Minary-Jolandan, M. Microscale 3D printing of nanotwinned copper. *Adv. Mater.* **2018**, *30*, 1705107. [CrossRef] [PubMed]
4. Seol, S.K.; Kim, D.; Lee, S.; Kim, J.H.; Chang, W.S.; Kim, J.T. Electrodeposition-based 3D printing of metallic microarchitectures with controlled internal structures. *Small* **2015**, *11*, 3896–3902. [CrossRef] [PubMed]
5. Francois, M.; Sun, A.; King, W.; Henson, N.; Tourret, D.; Bronkhorst, C.; Carlson, N.; Newman, C.; Haut, T.; Bakosi, J.; et al. Modeling of additive manufacturing processes for metals: Challenges and opportunities. *Curr. Opin. Solid State Mater. Sci.* **2017**, *21*, 198–206. [CrossRef]
6. Yang, Y.; Knol, M.; van Keulen, F.; Ayas, C. A semi-analytical thermal modelling approach for selective laser melting. *Addit. Manuf.* **2018**, *21*, 284–297. [CrossRef]
7. Yang, Y.; van Keulen, F.; Ayas, C. A computationally efficient thermal model for selective laser melting. *Addit. Manuf.* **2020**, *31*, 100955. [CrossRef]
8. Sundaram, M.M.; Kamaraj, A.B.; Kumar, V.S. Mask-less electrochemical additive manufacturing: A feasibility study. *J. Manuf. Sci. Eng.* **2015**, *137*, 021006. [CrossRef]
9. Madden, J.D.; Hunter, I.W. Three-dimensional microfabrication by localized electrochemical deposition. *J. Microelectromechanical Syst.* **1996**, *5*, 24–32. [CrossRef]
10. Said, R. Localized electro-deposition (LED): The march toward process development. *Nanotechnology* **2004**, *15*, S649. [CrossRef]
11. Je, J.H.; Kim, J.M.; Jaworski, J. Progression in the fountain pen approach: From 2D writing to 3D free-form micro/nanofabrication. *Small* **2017**, *13*, 1600137. [CrossRef]
12. Hu, J.; Yu, M.F. Meniscus-confined three-dimensional electrodeposition for direct writing of wire bonds. *Science* **2010**, *329*, 313–316. [CrossRef] [PubMed]
13. Kim, K.H.; Moldovan, N.; Espinosa, H.D. A nanofountain probe with sub-100 nm molecular writing resolution. *Small* **2005**, *1*, 632–635. [CrossRef] [PubMed]
14. Kim, K.; Ke, C.; Moldovan, N.; Espinosa, H. Massively parallel multi-tip nanoscale writer with fluidic capabilities-fountain pen nanolithography (FPN). In Proceedings of the 4th International Symposium on MEMS and Nanotechnology, Charlotte, NC, USA, 2–4 June 2003; pp. 235–238.
15. Wu, B.; Ho, A.; Moldovan, N.; Espinosa, H.D. Direct deposition and assembly of gold colloidal particles using a nanofountain probe. *Langmuir* **2007**, *23*, 9120–9123. [CrossRef] [PubMed]
16. Loh, O.; Lam, R.; Chen, M.; Moldovan, N.; Huang, H.; Ho, D.; Espinosa, H.D. Nanofountain-probe-based high-resolution patterning and single-cell injection of functionalized nanodiamonds. *Small* **2009**, *5*, 1667–1674. [CrossRef]
17. Moldovan, N.; Kim, K.; Espinosa, H.D. A multi-ink linear array of nanofountain probes. *J. Micromech. Microeng.* **2006**, *16*, 1935. [CrossRef]
18. Hirt, L.; Grüter, R.R.; Berthelot, T.; Cornut, R.; Vörös, J.; Zambelli, T. Local surface modification via confined electrochemical deposition with FluidFM. *Rsc Adv.* **2015**, *5*, 84517–84522. [CrossRef]

19. Falola, B.D.; Suni, I.I. Low temperature electrochemical deposition of highly active elements. *Curr. Opin. Solid State Mater. Sci.* **2015**, *19*, 77–84. [CrossRef]
20. Reiser, A.; Lindén, M.; Rohner, P.; Marchand, A.; Galinski, H.; Sologubenko, A.S.; Wheeler, J.M.; Zenobi, R.; Poulikakos, D.; Spolenak, R. Multi-metal electrohydrodynamic redox 3D printing at the submicron scale. *Nat. Commun.* **2019**, *10*, 1–8. [CrossRef]
21. Chen, X.; Liu, X.; Ouyang, M.; Chen, J.; Taiwo, O.; Xia, Y.; Childs, P.R.; Brandon, N.P.; Wu, B. Multi-metal 4D printing with a desktop electrochemical 3D printer. *Sci. Rep.* **2019**, *9*, 3973. [CrossRef]
22. Ambrosi, A.; Webster, R.D.; Pumera, M. Electrochemically driven multi-material 3D-printing. *Appl. Mater. Today* **2020**, *18*, 100530. [CrossRef]
23. Wang, P.; Roberts, R.; Ngan, A. Direct microfabrication of oxide patterns by local electrodeposition of precisely positioned electrolyte: The case of Cu2O. *Sci. Rep.* **2016**, *6*, 1–9. [CrossRef] [PubMed]
24. Kim, H.; Kim, J.G.; Park, J.W.; Chu, C.N. Selective copper metallization of nonconductive materials using jet-circulating electrodeposition. *Precis. Eng.* **2018**, *51*, 153–159. [CrossRef]
25. Chen, X.; Liu, X.; Childs, P.; Brandon, N.; Wu, B. A low cost desktop electrochemical metal 3D printer. *Adv. Mater. Technol.* **2017**, *2*, 1700148. [CrossRef]
26. Juggurnath, D.; Dauhoo, M.; Elahee, M.; Khoodaruth, A.; Osowade, A.; Olakoyejo, O.; Obayopo, S.; Adelaja, A. Simulations of air-water two-phase flow in an inclined pipe. In Proceedings of the 13th International Conference on Heat Transfer, Fluid Mechanics and Thermodynamics, Portoroz, Slovenia, 17–19 July 2017; pp. 77–84.
27. Winquist, F.; Krantz-Rülcker, C.; Lundström, I. A miniaturized voltammetric electronic tongue. *Anal. Lett.* **2008**, *41*, 917–924. [CrossRef]
28. Hoshyargar, V.; Nezameddin Ashrafizadeh, S.; Sadeghi, A. Diffusioosmotic flow in rectangular microchannels. *Electrophoresis* **2016**, *37*, 809–817. [CrossRef]
29. Lu, L.; Chen, X.; Huang, X.; Lu, K. Revealing the maximum strength in nanotwinned copper. *Science* **2009**, *323*, 607–610. [CrossRef]
30. Yeung, J.; Keong, L.C. Hardness measurement of copper bonding wire. *Procedia Eng.* **2014**, *75*, 134–139. [CrossRef]
31. Chandrasekar, M.; Pushpavanam, M. Pulse and pulse reverse plating—Conceptual, advantages and applications. *Electrochim. Acta* **2008**, *53*, 3313–3322. [CrossRef]
32. Pan, Y.; Liu, Y.; Wang, T.; Lu, X. Effect of a Cu seed layer on electroplated Cu film. *Microelectron. Eng.* **2013**, *105*, 18–24. [CrossRef]
33. Moharrami, N.; Bull, S. A comparison of nanoindentation pile-up in bulk materials and thin films. *Thin Solid Film.* **2014**, *572*, 189–199. [CrossRef]

crystals

MDPI

Article

Analytical Modeling of Residual Stress in Laser Powder Bed Fusion Considering Part's Boundary Condition

Elham Mirkoohi [1,*], Hong-Chuong Tran [2], Yu-Lung Lo [2], You-Cheng Chang [2], Hung-Yu Lin [2] and Steven Y. Liang [1]

1 Woodruff School of Mechanical Engineering, Georgia Institute of Technology, Atlanta, GA 30332, USA; steven.liang@me.gatech.edu
2 Department of Mechanical Engineering, National Cheng Kung University, Tainan City 701, Taiwan; hongchuong3389@gmail.com (H.-C.T.); loyl@mail.ncku.edu.tw (Y.-L.L.); n16074425@mail.ncku.edu.tw (Y.-C.C.); n16074409@mail.ncku.edu.tw (H.-Y.L.)
* Correspondence: elham.mirkoohi@gatech.edu

Received: 31 March 2020; Accepted: 21 April 2020; Published: 24 April 2020

Abstract: Rapid and accurate prediction of residual stress in metal additive manufacturing processes is of great importance to guarantee the quality of the fabricated part to be used in a mission-critical application in the aerospace, automotive, and medical industries. Experimentations and numerical modeling of residual stress however are valuable but expensive and time-consuming. Thus, a fully coupled thermomechanical analytical model is proposed to predict residual stress of the additively manufactured parts rapidly and accurately. A moving point heat source approach is used to predict the temperature field by considering the effects of scan strategies, heat loss at part's boundaries, and energy needed for solid-state phase transformation. Due to the high-temperature gradient in this process, the part experiences a high amount of thermal stress which may exceed the yield strength of the material. The thermal stress is obtained using Green's function of stresses due to the point body load. The Johnson–Cook flow stress model is used to predict the yield surface of the part under repeated heating and cooling. As a result of the cyclic heating and cooling and the fact that the material is yielded, the residual stress build-up is precited using incremental plasticity and kinematic hardening behavior of the metal according to the property of volume invariance in plastic deformation in coupling with the equilibrium and compatibility conditions. Experimental measurement of residual stress was conducted using X-ray diffraction on the fabricated IN718 built via laser powder bed fusion to validate the proposed model.

Keywords: residual stress prediction; IN718; experimental measurement of residual stress; additive manufacturing

---

1. Introduction

Laser powder bed fusion (L-PBF) is a type of metal additive manufacturing (AM) process that produces metallic parts layer by layer via selectively melting the powders. In the past few years, L-PBF is utilized to manufacture a wide variety of parts and assemblies using a vast range of material systems [1]. Laser powder bed fusion process has several superiorities over conventional manufacturing processes including reduction in density, fabrication of more intricate parts than has been previously possible, a reduction in design and manufacturing time due to the single-step manufacturing, fabrication of functionality graded material, and many more [2]. Nevertheless, parts built via L-BPF usually contain a high level of residual stress due to the large temperature gradient and cooling rates [3]. This may cause the part to fail during or after fabrication due to crack initiation and growth as caused by large stress [4,5]. Components produced via L-PBF are used in aerospace, automotive, medical, and marine industries; thus, high-quality components should be fabricated to be used in these critical applications [1,6]. An essential part of this qualification is the ability to accurately and rapidly predict the stress state within the part [7,8].

There is significant work in the literature on understanding of the residual stress formation during metal AM process under different process conditions [9]. This line of research can be categorized into three main parts: experimental measurements, numerical modeling, and analytical modeling.

Residual stress measurements can be classified into destructive and non-destructive methods [10]. The popular non-destructive methods are X-ray diffraction and Neutron diffraction techniques, which are capable of surface and volumetric residual stress measurements, respectively [11,12]. Destructive methods such as hole drilling, sectioning, and crack compliance mainly involved creating free surfaces and relating the deformation to the resultant residual stresses. Wu et al. [13] measured the residual stress using digital image correlation in conjunction with build plate removal and sectioning. The results are compared to the nondestructive volumetric neutron diffraction technique. They concluded that the residual stress is reduced by decreasing the island size, increasing the applied energy per length, and increasing island to wall rotation to 45°. Staub et al. [14] measured the residual stress using the X-ray technique in the L-PBF of SS316L. They concluded that the higher aspect ratio (width/depth) of the melt pool geometry could result in higher residual stress. Wang et al. [15] used X-ray method to measure the residual stress from electron beam additive manufacturing (EBAM) process as well as selective laser melting (SLM) for Ti-6Al-4V and IN718 parts. They summarized that the residual stress of Ti-6Al-4V parts that are made using EBAM is more compressive compared to IN-718. In addition, Ti-6Al-4V parts have a lower absolute value of residual stress compared to IN718 parts that are built using EBAM.

As explained above, there is significant research on the understanding of the residual stress formation under different process conditions using experimentations and numerical modeling. Experimental techniques to measure the residual stress within the part built via L-PBF certainly play an important role in the understanding of the residual stress formation. However, experimental measurement of the stress state in an entire part is time-consuming and expensive. Finite element models could predict the residual stress build-up in the AM components. However, performing FE simulations is challenged by the fact that it is computationally expensive. Consequently, many simplifications in modeling are conducted by applying different methods to predict the residual stress. While these simplifications are applied to this line of modeling, it still needs high-processor computers for the modeling of the entire process.

Numerical modeling is another approach for the prediction of residual stress in metal AM. Hajializade et al. [16] performed a finite element (FE) simulation using an adaptive mesh coarsening algorithm to increase the computational efficiency of the FE model for the direct metal deposition (DMD) of the AISI 304L. They used the double elliptical heat source model to predict the temperature distribution and used the result of thermal modeling to predict the residual stress. The heat loss due to the radiation is ignored in this modeling. Cheng et al. [17] performed FE modeling to predict the residual stress to design the support structure to mitigate residual stress formation. In this modeling, they used the inherent strain method to decrease the computational time. Siewert et al. [18] predicted the residual stress by using the Mechanical Layer Equivalent (MLE) method to decrease the computational time. They validated this model by measuring the residual stress using X-ray.

Herein, a novel physics-based analytical model is proposed to predict the residual stress in the IN718 parts built via L-PBF. The thermal signature of this process is predicted using a moving heat source approach by considering the scan strategy of the hatching space, layer thickness, and scan path, energy needed for solid-state phase change, and heat loss due to the convection and radiation. The high-temperature gradient and non-uniform heating in this process impose thermal stress on the component. The thermal stress is obtained by combining three sources of stresses known as stress due to the body forces, normal tension, and hydrostatic stress. Thermal stress may exceed the yield strength of the part. Hence, the Johnson–Cook (J-C) fellow stress model is used to predict the yield surface. As a result of repeated heating and cooling and the fact that the material is yielded, the through-thickness residual stress along the scan direction and build direction is predicted using the incremental plasticity and kinematic hardening behavior of the metal according to the property of volume invariance in plastic deformation in coupling with equilibrium and compatibility conditions. The X-ray diffraction technique was used to measure the residual stress of the IN718 components built via L-PBF to validate the proposed model. Good agreement was obtained between predicted and measured residual stress, which indicates that the proposed model is a valuable tool for the rapid and accurate prediction of the residual stress build-up in the parts built via L-PBF. Due to the high computational efficiency of the proposed model, it can also be used for the real-time monitoring and control of the build process, as well as for optimization of the process parameters in achieving a high-quality part.

2. Process Modeling

A physics-based closed-form analytical thermo-mechanical model is proposed to predict the residual stress of the IN718 parts built via L-PBF. The model significantly reduces the computational time, which makes it a great tool for the prediction, control, and optimization of the L-PBF process to achieve a high-quality part by cutting the trial-and-error approach.

2.1. Thermal Modeling

A moving point heat source approach is used to predict the temperature field within the parts built via L-PBF. The initial solution of temperature in a semi-infinite medium was derived by Carslaw and Jaeger [19]. Heat losses at part's boundaries including conduction on lateral faces and convection and radiation on the top surface are solved by modifying the heat source solution. Figure 1 illustrates the heat transfer mechanisms in L-PBF. The closed-form solution of temperature considering part's boundary condition can be obtained as

$$T = \frac{1}{4\pi KR(T-T_0)} \times \left\{ P\exp\left(\frac{-V(R+x)}{2D}\right) - A\left[h(T-T_0) + \varepsilon\sigma\left(T^4 - T_0^4\right) + \frac{K(T-T_0)}{R}\right]\right\} + T_0 \quad (1)$$

where P is the laser power, α represents absorption coefficient, K is the thermal conductivity, $R = \sqrt{x^2+y^2+z^2}$ is the radial distance from heat source, V is the laser speed, $D = \frac{K}{\rho C}$ is the thermal diffusivity, ρ is the density, C is the specific heat, A is the area of the heat sink, h is convection coefficient, ε is the emissivity, σ is the Stefan–Boltzmann constant, n is the number of heat sinks, i is the index of each heat sink, and T_0 is the initial temperature. More explanation about the heat source solution can be obtained from the previous work of the authors [20–23].

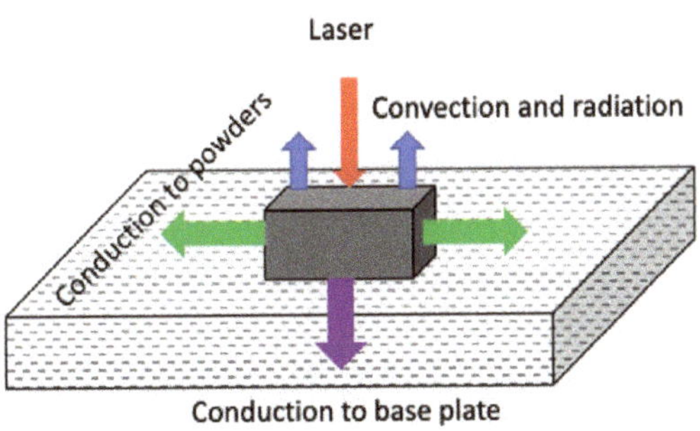

Figure 1. Heat transfer mechanisms in laser powder bed fusion.

2.2. Thermal Stress Prediction

Thermal stress induced by non-uniform heating is obtained using Green's function of stresses due to the point body load [23,24]. The obtained thermal stress is the combination of stresses due to the body forces, normal tension, and hydrostatic stress, as described by Saif et al. [25]:

$$\begin{aligned}\sigma_{xx}(x,z) = &-\frac{\alpha(T)E(T)}{1-2\nu(T)}\int_0^{\infty}\int_{-\infty}^{\infty}(G_{xh}\frac{\partial T}{\partial x}(x',z') + G_{xv}\frac{\partial T}{\partial z}(x',z'))dx'dz' \\ &+\frac{2z}{\pi}\int_{-\infty}^{\infty}\frac{p(s)(s-x)^2}{((s-x)^2+z^2)^2}ds - \frac{\alpha(T)E(T)T(x,z)}{1-2\nu(T)}\end{aligned} \tag{2}$$

$$\begin{aligned}\sigma_{zz}(x,z) = &-\frac{\alpha(T)E(T)}{1-2\nu(T)}\int_0^{\infty}\int_{-\infty}^{\infty}(G_{zh}\frac{\partial T}{\partial x}(x',z') + G_{zv}\frac{\partial T}{\partial z}(x',z'))dx'dz' \\ &+\frac{2z^3}{\pi}\int_{-\infty}^{\infty}\frac{p(s)}{((s-x)^2+z^2)^2}ds - \frac{\alpha(T)E(T)T(x,z)}{1-2\nu(T)}\end{aligned} \tag{3}$$

$$\sigma_{xz}(x,z) = -\frac{\alpha(T)E(T)}{1-2\nu(T)}\int_0^{\infty}\int_{-\infty}^{\infty}(G_{xzh}\frac{\partial T}{\partial x}(x',z') + G_{xzv}\frac{\partial T}{\partial z}(x',z'))dx'dz' + \frac{2z^2}{\pi}\int_{-\infty}^{\infty}\frac{p(s)(s-x)}{((s-x)^2+z^2)^2}ds \tag{4}$$

$$\sigma_{yy}(x,z) = \nu(T)(\sigma_{xx}+\sigma_{zz}) - \alpha(T)E(T)T(x,z) \tag{5}$$

where α represents the coefficient of the thermal expansion, E is the elastic modulus, $\frac{\partial T}{\partial x}$ is the temperature gradient and $p(s)$ is expressed by:

$$p(s) = \frac{\alpha(T)E(T)T(x,z=0)}{1-2\nu(T)} \tag{6}$$

G_{xh}, G_{xV}, G_{zh}, G_{zv}, G_{xzh}, G_{xzv} are the plane strain Green's functions, which are explained in [26].

The closed-form solution of thermal stress can be derived as

$$\sigma_{xx}(i,j) = -\frac{\alpha E(T)}{1-2\nu}(G_{xh}T_x + G_{xv}T_z)\left(|x_i - x_{i-1}||z_j - z_{j-1}|\right) + \frac{\frac{2z}{\pi}(p(s)(s-x)^2}{\left((s-x)^2+z^2\right)^2}|x_i - x_{i-1}| - \frac{\alpha ET}{1-2\nu} \tag{7}$$

where $T_x = \frac{T(i,j)-T(i-1,j)}{|x_i - x_{i-1}|}$, $T_z = \frac{T(i,j)-T(i,j-1)}{|z_j - z_{j-1}|}$ and i and j represent a location in the 2D medium at which the stress is calculated. The closed-form solutions of σ_{zz} and σ_{xz} can be derived using the same method.

2.3. Residual Stress Prediction

The Johnson–Cook (J-C) flow stress model is employed to determine the yield surface of the material as

$$\sigma = \left(A + B\varepsilon_{eff}^{p\;n}\right)\left(1 + Cln\left(\frac{\dot{\varepsilon}_{eff}^{p}}{\dot{\varepsilon}_0}\right)\right)\left(1 - \left[\frac{T-T_0}{T_m - T_0}\right]^m\right) \tag{8}$$

where ε_{eff}^{p} is the effective plastic strain, $\dot{\varepsilon}_{eff}^{p}$ is the effective plastic strain rate, T is the temperature of material, T_m is the melting point of material, and T_0 is the initial temperature. The terms A, B, C, n, m, and $\dot{\varepsilon}_0$ are the material constant for IN718, which are listed in Table 1.

The yielding criterion is obtained for an isotropic material. Kinematic hardening is considered by employing backstress tensor (α_{ij})

$$F_{yeild} = \frac{3}{2}\left(S_{ij} - \alpha_{ij}\right)\left(S_{ij} - \alpha_{ij}\right) - k^2 = 0 \tag{9}$$

where $S_{ij} = \sigma_{ij} - \left(\frac{\sigma_{kk}}{3}\right)\delta_{ij}$ is the deviatoric stress.

If $F_{yeild} < 0$, the material is in elastic region and the stresses can be obtained from the Hook's Law. If $F_{yeild} > 0$, incremental plastic strains are calculated and accumulated during the stress history to determine the total plastic strains.

In the elastic–plastic case where the $F_{yeild} \geq 0$, the strain rate along the scan direction and transverse direction can be calculated using modified McDowell algorithm.

$$\begin{cases} \frac{1}{E}\left[\dot{\sigma}_{xx} - v\left(\dot{\sigma}_{yy} - \dot{\sigma}_{zz}\right)\right] + \alpha\Delta T + \frac{1}{h}\left(\dot{\sigma}_{xx}n_{xx} + \dot{\sigma}_{yy}n_{yy} + \dot{\sigma}_{zz}n_{zz} + 2\dot{\sigma}^{*}_{xz}n_{xz}\right)n_{xx} = \\ \psi\left(\frac{1}{E}\left[\dot{\sigma}^{*}_{xx} - v\left(\dot{\sigma}_{yy} - \dot{\sigma}^{*}_{zz}\right)\right] + \alpha\Delta T + \frac{1}{h}\left(\dot{\sigma}^{*}_{xx}n_{xx} + \dot{\sigma}_{yy}n_{yy} + \dot{\sigma}^{*}_{zz}n_{zz} + 2\dot{\sigma}^{*}_{xz}n_{xz}\right)n_{xx}\right) \\ \frac{1}{E}\left[\dot{\sigma}_{yy} - v\left(\dot{\sigma}_{xx} - \dot{\sigma}_{zz}\right)\right] + \alpha\Delta T + \frac{1}{h}\left(\dot{\sigma}_{xx}n_{xx} + \dot{\sigma}_{yy}n_{yy} + \dot{\sigma}_{zz}n_{zz} + 2\dot{\sigma}^{*}_{xz}n_{xz}\right)n_{yy} = 0 \\ \dot{\sigma}_{yy} = \frac{1}{2}\left(\dot{\sigma}_{xx} + \dot{\sigma}_{zz}\right) \end{cases} \tag{10}$$

where $\dot{\sigma}^{*}_{xx}$, $\dot{\sigma}^{*}_{zz}$, $\dot{\sigma}^{*}_{xz}$ are the elastic thermal stresses calculated from Equations (2) and (6). In the McDowell model, a hybrid function (ψ) is proposed, which depends on the instantaneous value of the modulus ratio h/G as

$$\psi = 1 - \exp\left(-\xi\frac{3h}{2G}\right) \tag{11}$$

where $\xi = 0.15$ is the algorithm constant, h is the plastic modulus, and $G = E/(2(1+v))$ is the elastic shear modulus. Three systems of equations are solved simultaneously for $\dot{\sigma}_{xx}$, $\dot{\sigma}_{yy}$, and $\dot{\sigma}_{zz}$ for each elastic–plastic increment of strain [27].

Table 1. Johnson–Cook parameters for IN718 [28].

A (MPa)	*B* (MPa)	*C*	*n*	*m*	$\dot{\varepsilon}_0$
980	1370	0.02	0.164	1.03	1

After one layer is laser scanned, elastic stresses are relaxed to satisfy the boundary condition prescribed by Merwin and Johnson [29] as

$$\begin{gathered} \varepsilon^{r}_{xx} = 0, \quad \sigma^{r}_{xx} = f_1(z), \quad \varepsilon^{r}_{yy} = 0, \quad \sigma^{r}_{yy} = f_2(z), \quad \varepsilon^{r}_{zz} = f_3(z), \quad \sigma^{r}_{zz} = 0, \\ \gamma^{r}_{xz} = f_4(z), \quad \sigma^{r}_{xz} = 0 \end{gathered} \tag{12}$$

Finally, only stresses and strains parallel to the surface (σ^{r}_{xx}, σ^{r}_{yy}, γ^{r}_{xz}) remain non-zero [30].

The proposed analytical model does not consider the effect of geometry. Moreover, the bulk temperature dependent material properties are considered for the simulation of residual stress. It should be noted that the proposed model can be used for both powder bed and powder feed systems by just changing the boundary conditions in the temperature modeling. In powder feed systems, since the part is subjected to air, the heat loss from lateral faces would be due to convection.

3. Experimental Procedure

3.1. Build Process

Density of the additively manufactured part built via L-PBF has a crucial impact on mechanical properties of the fabricated component. An approach to identify processing parameters for producing high-density parts was employed to select the processing conditions, as described in the previous studies [31–34]. Tong Tai AM250 selective laser melting (SLM) machine (Kaohsiung, Taiwan), equipped with a 50–400 W YAG laser with the laser spot size of D4sigma = 54 μm, was used to fabricate rectangular bar shape samples with dimensions of 10 × 5 × 5 mm^3. The processing conditions used to fabricate high-density samples for measuring the residual stress are listed in Table 2. The spherical IN718 powders used in this study are distributed in the range of d_{10} = 17.51 μm, d_{50} = 31.44 μm, and d_{90} = 52.21 μm.

Table 2. Process parameters designed for the fabrication of IN718 specimens using L-PBF.

Sample	Laser Power (W)	Scan Speed (mm/s)	Layer Thickness (μm)	Hatching Space (μm)	Number of Scans	Pre-Heating	Rotation Angle of Scan Vector between Layers (°)
1	150	600	30	100	50	No	67
2	250	600	30	100	50	No	67
3	150	1000	30	100	50	No	67

The process parameters were selected in a way to capture the effects of scan speed and laser power on residual stress; thus, the other parameters of layer thickness, hatching space, and hatch style were kept the same for all samples. Figure 2 illustrates the as-built samples and were then removed from the base plate using the electrical discharge machining (EDM).

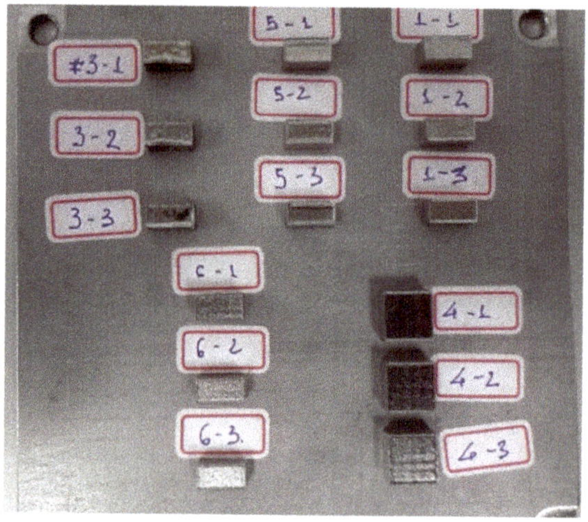

Figure 2. Illustration of the as-built IN718 parts manufactured via L-PBF.

3.2. *Experimental Measurement of Residual Stress*

The residual stress on the side walls of the samples, as illustrated in Figure 3, was measured by a commercial X-ray Diffractions machine (D8 Discover Bruker, Billerica, MA, USA) using the $sin^2\Psi$ method [35,36]. The coordinate and location of measured points are shown in Table 3. For each point shown in Figure 3, the through-thickness residual stresses along the build direction (z-direction) and along the scan direction (x-direction) were measured. The parameters for XRD measurement are specified in Table 4. It should be noted that, for each set of parameters and each point shown in Figure 3, the XRD measurements were performed on three fabricated samples and the results were averaged.

Table 3. Coordinates of the measured points using XRD.

A	x = 5 mm; y = 0; z = 0.5 mm
B	x = 5 mm; y = 0; z = 1.5 mm
C	x = 5 mm; y = 0; z = 2.5 mm
D	x = 5 mm; y = 0; z = 3.5 mm
E	x = 5 mm; y = 0; z = 4.5 mm

Table 4. Parameters for XRD measurements.

Focus	1.0 mm
Radiation	Cu Kα
Lattice Plane (hkl) [36]	{420}
2θ [36]	145°
Ψ-Tilting	0° to 45° in 6 steps each
Young Modulus [37]	199,955 MPa
Poisson Ration [37]	0.29

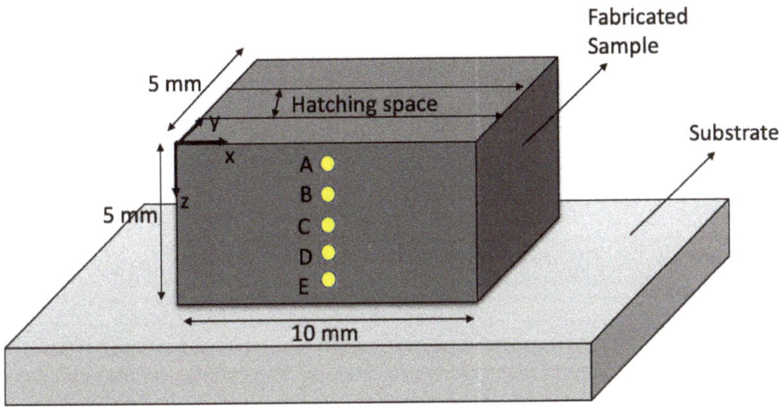

Figure 3. Schematic diagram illustrating the definition of processing parameters.

4. Results and Discussion

The proposed thermomechanical analytical model enables the prediction of the residual stress of the fabricated L-PBF samples under different process conditions along the scan direction (σ_{xx}) and build direction (σ_{zz}). The thermomechanical material properties of the IN718 are listed in Table 5. The thermal response of the build part was predicted using moving point heat source approach by considering the effects of heat loss due to the convection and radiation, scan strategies, and energy needed for solid-state phase transformation. Figures 4–6 illustrate the predicted temperature field for three IN718 samples listed in Table 5. The analytical modeling of the temperature field is validated in the previous work of authors [20].

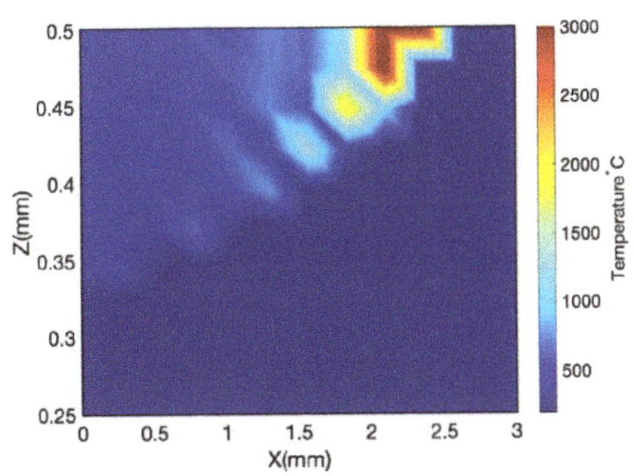

Figure 4. Predicted temperature field for the fabricated IN718 sample built via L-PBF with the laser power of 150 W, scan speed of 600 mm/s, layer thickness of 30 μm, and hatching space of 100 μm.

Table 5. Temperature dependent material properties of IN718 (temperature is in °C) [37–41].

Density g/cm^3	
$\rho = 8.19 - 39.2 \times 10^{-2}T$	$25 < T \leq 1170$
$\rho = 7.40 - 88.0 \times 10^{-2}(T - 1200)$	$T > 1170$
Thermal Conductivity W/m°C	
$k = 39.73 - 24.0 \times 10^{-3}T + 2 \times 10^{-3}T^{-2}$	$25 < T < 1170$
$k = 29.6$	$T > 1170$
Specific Heat J/kg°C	
$C_p = 420.24 + 0.026T - 4 \times 10^{-6}T^2$	$25 < T \leq 1170$
$C_p = 650$	$T > 1170$
Thermal Expansion 1/°C	
$\alpha = -9 \times 10^{-13}T^2 - 7.7 \times 10^{-9}T + 1.1 \times 10^{-5}$	$25 < T \leq 1100$
$\alpha = 1.8 \times 10^{-5}$	$T > 1100$
Elastic Modulus GPa	
$E = 5.2 \times 10^{-5}T^2 - 0.088T + 1.6 \times 10^2$	$25 < T \leq 798$
$E = 3.1 \times 10^{-5}T^2 - 0.23T + 2.9 \times 10^2$	$798 < T < 2500$
Yield Strength MPa	
$\sigma_Y = -9 \times 10^{-10}T^4 - 1.2 \times 10^{-6}T^3 + 0.00026T^2 - 0.23T + 3.2 \times 10^2$	$25 < T < 2500$
Poisson's Ratio	
$\nu = -4.8 \times 10^{-10}T^3 - 8.8 \times 10^{-7}T^2 - 0.00031T + 0.31$	$25 < T < 2500$

Figure 4 demonstrates the predicted temperature field for the first sample with the laser power of 150 W, scan speed of 600 mm/s, layer thickness of 30 µm, and hatching space of 100 µm. It should be noted that the simulated dimension is 5 mm × 0.5 mm × 0.5 mm. The laser was located at the middle of the samples to simulate the temperature field just at the irradiation of the laser. This location of the laser was selected since the experimental measurements were gathered from the middle of the samples to be consistent.

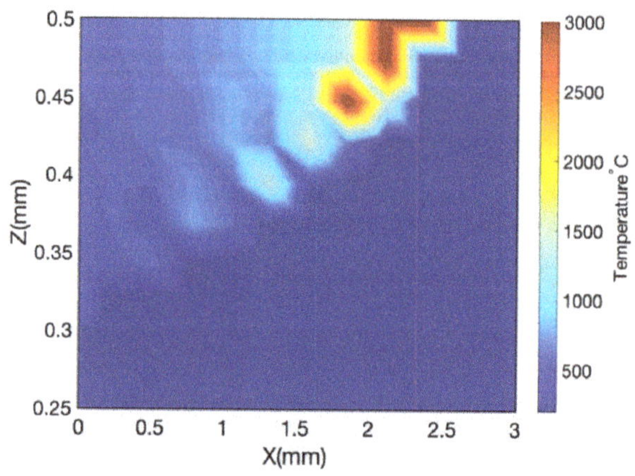

Figure 5. Predicted temperature field for the fabricated IN718 sample built via L-PBF with the laser power of 250 W, scan speed of 600 mm/s, layer thickness of 30 µm, and hatching space of 100 µm.

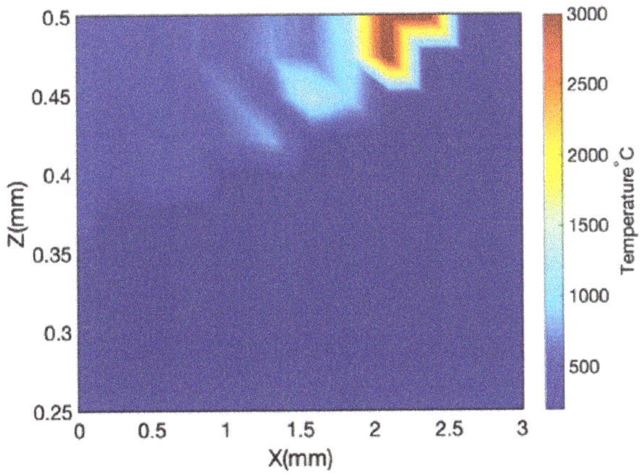

Figure 6. Predicted temperature field for the fabricated IN718 sample built via L-PBF with the laser power of 150 W, scan speed of 1000 mm/s, layer thickness of 30 μm, and hatching space of 100 μm.

Figure 5 illustrates the predicted temperature field for the second sample with the laser power of 250 W, scan speed of 600 mm/s, layer thickness of 30 μm, and hatching space of 100 μm. Figure 6 depicts the predicted temperature field for the third sample with the laser power of 150 W, scan speed of 1000 mm/s, layer thickness of 30 μm, and hatching space of 100 μm.

Comparison of Figures 4 and 5 shows that the increase in laser power would increase the melt pool area, and the comparison of Figures 4 and 6 illustrates that, for the same laser power, the increase in scan speed would decrease the melt pool geometry, which is very well-known knowledge and confirms that the proposed model properly follows the trend. The proposed analytical model of the temperature field is validated in the previous work of these authors.

High-temperature gradient and fast cooling rate induce thermal stress in the fabricated part. The thermal stress is calculated using Green's function of stresses due to the point body load. The thermal stress is a combination of three main sources of stress known as stresses due to body forces, normal tension, and hydrostatic stress. The thermal stress may exceed the yield strength of the material. The yield surface is predicted using the J-C flow stress model. Due to the repeated heating and cooling and the fact that the material is yielded, the part experiences a high level of residual stress. Residual stress is calculated using incremental plasticity and kinematic hardening behavior of the material in coupling with equilibrium and compatibility conditions.

As shown in Figure 3, the residual stress was measured at five different locations through the thickness along the scan direction and build direction. The residual stress was predicted for two different heat transfer boundary conditions: (1) the residual stress considering the heat transfer boundary conditions; and (2) the residual stress considering heat transfer boundary condition. Then, both conditions were compared to experimental measurements.

Figure 7a,b illustrates the predict residual stress for the IN718 sample, which is built with the laser power of 150 W, scan speed of 600 mm/s, layer thickness of 30 μm, and hatching space of 100 μm. The predicted and measured residual stresses are highly tensile along both scan direction and build direction since, during the cooling cycle, the heat-affected zone begins to cool down and the shrinkage of material in this zone tends to occur; thus, the tensile stress state builds up in the heated zone. Moreover, the predicted and measured residual stress along the build direction showed higher values compared to that along the scan direction, which is due to the fact that the heat transfer mechanism varies in different directions and would result in different temperature gradient and cooling rates [38]. The variation in temperature gradient and cooling rate resulting from different heat transfer mechanisms would impact the microstructural evolution and material properties, which then impact the residual stress. Furthermore, the comparison of the predicted residual stress with and without considering the effect of heat loss reveals that, although in most of the points through thickness the predicted residual stress in both conditions are within the range of experimental measurements, the predicted residual stress considering the effect of heat loss demonstrates lower bound predictions. This is because, due to the heat loss, the temperature gradient would be lower compared to that without the effect of heat loss. It should also be noted that the predicted residual stress near the surface for these two conditions show more divergence compared to that at higher depth into the workpiece. This is because, the heat loss has higher impact on the heat affected zone and lower impact on the solidified material. Thus, the predictions are more sensitive to the heat loss near the surface where the heat affected zone exists.

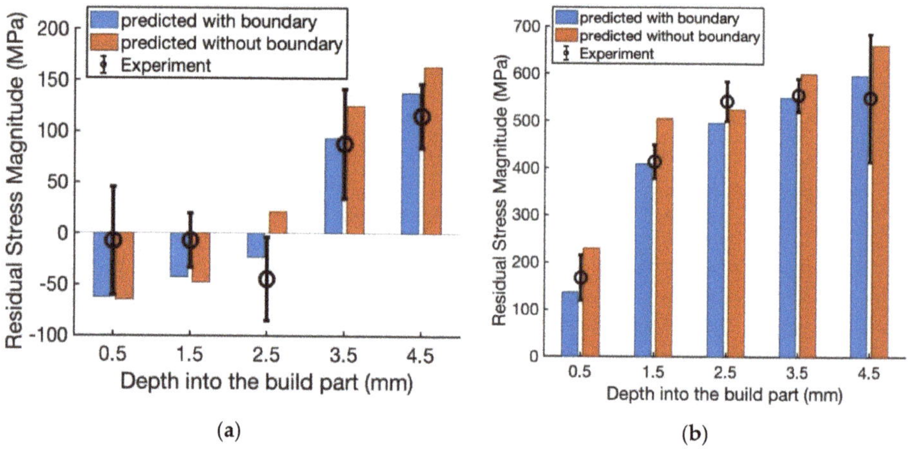

Figure 7. Predicted and measured residual stress of IN718 sample built via L-PBF with the laser power of 150 W and scan speed of 600 mm/s, layer thickness of 30 µm, and hatching space of 100 µm: (**a**) along the scan direction; and (**b**) along the build direction.

Figure 8a,b demonstrates the predicted and measured residual stress of IN718 sample with the laser power of 250 W, scan speed of 600 mm/s, layer thickness of 30 µm, and hatching space of 100 µm. The predicted residual stress along the scan direction and build direction are highly tensile in accordance with the data in the literature. Furthermore, the predicted residual stress along the scan direction has lower value compared to that along the build direction. As explained above, different heat transfer mechanisms contribute to this variation. Moreover, the predicted residual stress without considering the effect of heat loss demonstrates upper bound predictions. This is because the temperature gradient is higher compared to that when the effect of heat loss is considered, thus resulting in higher prediction of residual stress values.

Figure 9a,b shows the predicted and measured residual stress along the scan direction and build direction for the laser power of 150 W, scan speed of 1000 mm/s, layer thickness of 30 µm, and hatching space of 100 µm. As shown in these figures, the predicted residual stress is highly tensile along the scan direction and build direction. Moreover, the predicted residual stress without considering the effect of heat loss demonstrate higher values compared to that when the effect of heat loss is considered, as explained above. The predicted residual stress showed good agreement with measured residual stress of IN718.

To sum up, the comparison of the predicted and measured residual stress for three IN718 samples showed that the proposed thermo-mechanical analytical is a valuable, reliable, and rapid tool for the prediction of stress state within the part which then can be used for the optimization of process parameters and control of the build process through inverse analysis [39–41]. It should be noted that the proposed model works for all material systems as long as the thermo-mechanical material properties are known.

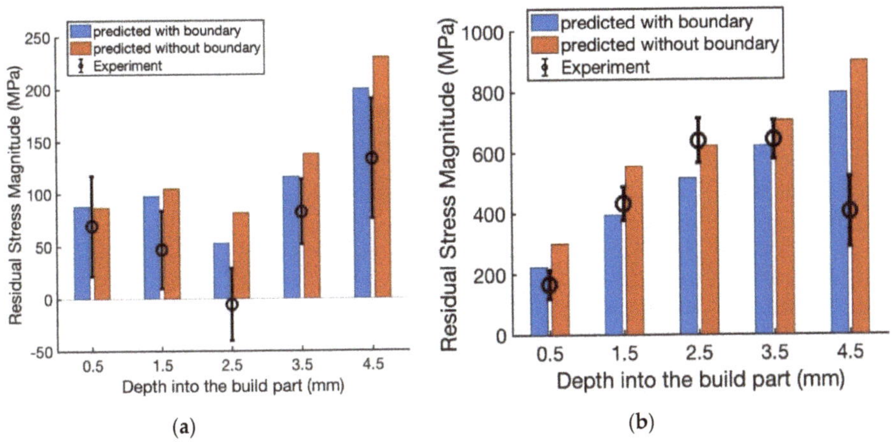

Figure 8. Predicted and measured residual stress of IN718 sample built via L-PBF with the laser power of 250 W and scan speed of 600 mm/s, layer thickness of 30 μm, and hatching space of 100 μm: (**a**) along the scan direction; and (**b**) along the build direction.

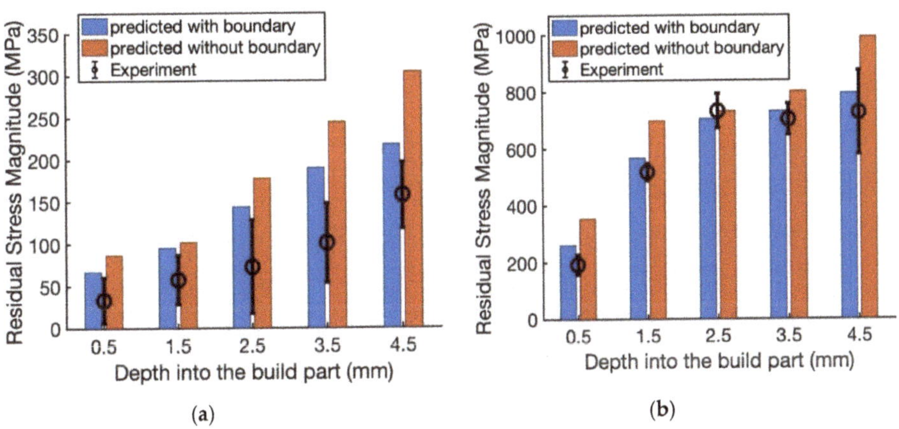

Figure 9. Predicted and measured residual stress of IN718 sample built via L-PBF with the laser power of 150 W and scan speed of 1000 mm/s, layer thickness of 30 μm, and hatching space of 100 μm: (**a**) along the scan direction; and (**b**) along the build direction.

5. Conclusions

A physics-based analytical model is proposed to predict the residual stress in laser powder bed fusion process. The proposed model was validated by measuring the residual stress of the fabricated IN718 samples via X-ray diffraction. The proposed thermo-mechanical analytical model predicts the temperature field of the additively manufactured part using moving point heat source approach by considering the effects of scan strategies, heat loss due to convection and radiation, and energy needed for solid-state phase change. The build part may experience high thermal stress due to the high-temperature gradient. The thermal stress is obtained using Green's function of stresses due to the point body load. The thermal stress may exceed the yield strength of the material. Thus, the Johnson–Cook flow stress model is used to determine the yield surface. Due to the cyclic heating and cooling, the material experiences high residual stress. The residual stress is predicted using incremental plasticity and kinematic hardening behavior of the metal according to the property of volume invariance in plastic deformation in coupling with the equilibrium and compatibility conditions.

The residual stress is predicted along the scan direction and build direction under three different process conditions. The results show that the residual stress is highly tensile in both directions since, during the cooling cycle, the heat-affected zone begins to cool down and the shrinkage of material in this zone tends to occur; thus, the tensile stress state builds up.

Moreover, a comparison of the residual stress along the scan direction and build direction showed that the predicted residual stress along the build direction is higher than that along the scan direction. Different heat transfer mechanisms along the scan direction and build direction could contribute to this difference.

Furthermore, the residual stress is predicted under two different boundary conditions: (1) the effect of heat loss due to the convection and radiation is considered; and (2) the effect of heat loss from boundaries is ignored. The results show that, when the heat loss is ignored, the predicted residual stress has higher value compared to the results where the effect of heat loss is considered. Although in most of the points through thickness, the results for both conditions are within the range of experimental measurements, the predicted residual stress without the effect of heat loss boundary conditions is an upper bound, and the predicted residual stress with the effect of heat loss boundary conditions is a lower bound.

Results from the proposed analytical residual stress model showed good agreement with X-ray diffraction measurements used to determine the residual stresses in the IN718 specimens built via L-PBF. Thus, the proposed model is a valuable tool for the rapid and accurate prediction of the residual stress build-up in the parts built via L-PBF. Due to the high computational efficiency of the proposed model, this model can also be used for the real-time monitoring and control of the build process, as well as optimization of the process parameters in achieving a high-quality part.

Author Contributions: Conceptualization, E.M.; methodology, E.M.; software, E.M.; validation, E.M., H.-C.T., Y.-L.L., Y.-C.C., H.-Y.L.; formal analysis, E.M.; investigation, E.M.; writing—original draft preparation, E.M.; writing—review and editing, S.Y.L.; visualization, E.M.; supervision, S.Y.L.; All authors have read and agreed to the published version of the manuscript.

Funding: This research received no external funding.

Acknowledgments: The experimental data were financially supported by the "Intelligent Manufacturing Research Center" (iMRC) from The Featured Areas Research Center Program within the framework of the Higher Education Sprout Project by the Ministry of Education (MOE) in Taiwan.

Conflicts of Interest: The authors declare no conflict of interest.

References

1. Bartlett, J.L.; Li, X. An overview of residual stresses in metal powder bed fusion. *Addit. Manuf.* **2019**, *27*, 131–149. [CrossRef]
2. King, W.E. Observation of keyhole-mode laser melting in laser powder-bed fusion additive manufacturing. *J. Mater. Process. Technol.* **2014**, *214*, 2915–2925. [CrossRef]
3. Khairallah, S.A. Laser powder-bed fusion additive manufacturing: Physics of complex melt flow and formation mechanisms of pores, spatter, and denudation zones. *Acta Mater.* **2016**, *108*, 36–45. [CrossRef]
4. Caiazzo, F. Laser powder-bed fusion of Inconel 718 to manufacture turbine blades. *Int. J. Adv. Manuf. Technol.* **2017**, *93*, 4023–4031. [CrossRef]
5. Masoomi, M.; Thompson, S.M.; Shamsaei, N. Laser powder bed fusion of Ti-6Al-4V parts: Thermal modeling and mechanical implications. *Int. J. Mach. Tools Manuf.* **2017**, *118*, 73–90. [CrossRef]
6. Levkulich, N. The effect of process parameters on residual stress evolution and distortion in the laser powder bed fusion of Ti-6Al-4V. *Addit. Manuf.* **2019**, *28*, 475–484. [CrossRef]
7. Barros, R. Laser Powder Bed Fusion of Inconel 718: Residual Stress Analysis Before and After Heat Treatment. *Metals* **2019**, *9*, 1290. [CrossRef]
8. Tabei, A.; Mirkoohi, E.; Garmestani, H.; Liang, S. Modeling of texture development in additive manufacturing of Ni-based superalloys. *Int. J. Adv. Manuf. Technol.* **2019**, *103*, 1057–1066. [CrossRef]
9. Bandyopadhyay, A.; Traxel, K.D. Invited review article: Metal-additive manufacturing—Modeling strategies for application-optimized designs. *Addit. Manuf.* **2018**, *22*, 758–774. [CrossRef]

10. An, K. Neutron residual stress measurement and numerical modeling in a curved thin-walled structure by laser powder bed fusion additive manufacturing. *Mater. Des.* **2017**, *135*, 122–132. [CrossRef]
11. Yadroitsev, I.; Yadroitsava, I. Evaluation of residual stress in stainless steel 316L and Ti6Al4V samples produced by selective laser melting. *Virtual Phys. Prototyp.* **2015**, *10*, 67–76. [CrossRef]
12. Liu, Y.; Yang, Y.; Wang, D. A study on the residual stress during selective laser melting (SLM) of metallic powder. *Int. J. Adv. Manuf. Technol.* **2016**, *87*, 647–656. [CrossRef]
13. Wu, A.S. An experimental investigation into additive manufacturing-induced residual stresses in 316L stainless steel. *Metall. Mater. Trans. A* **2014**, *45*, 6260–6270. [CrossRef]
14. Staub, A.; Spierings, A.B.; Wegener, K. Correlation of meltpool characteristics and residual stresses at high laser intensity for metal lpbf process. *Adv. Mater. Process. Technol.* **2019**, *5*, 153–161. [CrossRef]
15. Wang, Z. Residual stress mapping in Inconel 625 fabricated through additive manufacturing: Method for neutron diffraction measurements to validate thermomechanical model predictions. *Mater. Des.* **2017**, *113*, 169–177. [CrossRef]
16. Hajializadeh, F.; Ince, A. Finite element–based numerical modeling framework for additive manufacturing process. *Mater. Des. Process. Commun.* **2019**, *1*, e28. [CrossRef]
17. Cheng, B.; Chou, K. Geometric consideration of support structures in part overhang fabrications by electron beam additive manufacturing. *Comput. Aided Des.* **2015**, *69*, 102–111. [CrossRef]
18. Siewert, M. Validation of Mechanical Layer Equivalent Method for simulation of residual stresses in additive manufactured components. *Comput. Math. Appl.* **2019**, *78*, 2407–2416. [CrossRef]
19. Jaeger, J.C.; Carslaw, H.S. *Conduction of Heat in Solids*; Clarendon Press: Oxford, UK, 1959.
20. Ning, J. Analytical modeling of 3D temperature distribution in selective laser melting of Ti-6Al-4V considering part boundary conditions. *J. Manuf. Process.* **2019**, *44*, 319–326. [CrossRef]
21. Mirkoohi, E. Heat Source Modeling in Selective Laser Melting. *Materials* **2019**, *12*, 2052. [CrossRef]
22. Mirkoohi, E. Three-dimensional semi-elliptical modeling of melt pool geometry considering hatch spacing and time spacing in metal additive manufacturing. *J. Manuf. Process.* **2019**, *45*, 532–543. [CrossRef]
23. Mirkoohi, E. Thermal modeling of temperature distribution in metal additive manufacturing considering effects of build layers, latent heat, and temperature-sensitivity of material properties. *J. Manuf. Mater. Process.* **2018**, *2*, 63. [CrossRef]
24. Woo, W. Effect of interlayers and scanning strategies on through-thickness residual stress distributions in additive manufactured ferritic-austenitic steel structure. *Mater. Sci. Eng. A* **2019**, *744*, 618–629. [CrossRef]
25. Saif, M.; Hui, C.; Zehnder, A. Interface shear stresses induced by non-uniform heating of a film on a substrate. *Thin Solid Films* **1993**, *224*, 159–167. [CrossRef]
26. Mirkoohi, E. Thermo-mechanical modeling of thermal stress in metal additive manufacturing considering elastoplastic hardening. *CIRP J. Manuf. Sci. Technol.* **2020**, *28*, 52–67. [CrossRef]
27. McDowell, D. An approximate algorithm for elastic-plastic two-dimensional rolling/sliding contact. *Wear* **1997**, *211*, 237–246. [CrossRef]
28. Kobayashi, T. Plastic flow behavior of Inconel 718 under dynamic shear loads. *Int. J. Impact Eng.* **2008**, *35*, 389–396. [CrossRef]
29. Group, A.M.; Merwin, J.; Johnson, K. An analysis of plastic deformation in rolling contact. *Proc. Inst. Mech. Eng.* **1963**, *177*, 676–690. [CrossRef]
30. Mirkoohi, E.; Dobbs, J.R.; Liang, S.Y. Analytical modeling of residual stress in direct metal deposition considering scan strategy. *Int. J. Adv. Manuf. Technol.* **2020**, *106*, 4105–4121. [CrossRef]
31. Lo, Y.-L.; Tran, H.-C. Optimized Hatch Space Selection in Double-Scanning Track Selective Laser Melting Process. *Int. J. Adv. Manuf. Technol.* **2019**, *105*, 2989–3006. [CrossRef]
32. Tran, H.-C. Systematic Approach for Determining Optimal Processing Paramters to Produce Parts with High Density in Selective Laser Melting Process. *Int. J. Adv. Manuf. Technol.* **2019**, *105*, 4443–4460. [CrossRef]
33. Tran, H.-C.; Lo, Y.-L. Heat transfer simulations of selective laser melting process based on volumetric heat source with powder size consideration. *J. Mater. Process. Technol.* **2018**, *255*, 411–425. [CrossRef]
34. Tran, H.-C.; Lo, Y.-L.; Huang, M.-H. Analysis of Scattering and Absorption Characteristics of Metal Powder Layer for Selective Laser Sintering. *IEEE/ASME Trans. Mechatron.* **2017**, *22*, 1807–1817. [CrossRef]
35. Nadammal, N. Influence of support configurations on the characteristics of selective laser-melted inconel 718. *JOM* **2018**, *70*, 343–348. [CrossRef]
36. Prevey, P.S. X-ray diffraction residual stress techniques. *ASM Int.* **1986**, *10*, 380–392.

37. Yang, Y. Prediction of microstructure, residual stress, and deformation in laser powder bed fusion process. *Comput. Mech.* **2018**, *61*, 1–17. [CrossRef]
38. Ganeriwala, R. Evaluation of a thermomechanical model for prediction of residual stress during laser powder bed fusion of Ti-6Al-4V. *Addit. Manuf.* **2019**, *27*, 489–502. [CrossRef]
39. Mirkoohi, E.; Bocchini, P.; Liang, S.Y. An analytical modeling for process parameter planning in the machining of Ti-6Al-4V for force specifications using an inverse analysis. *Int. J. Adv. Manuf. Tech.* **2018**, *98*, 2347–2355. [CrossRef]
40. Mirkoohi, E.; Bocchini, P.; Liang, S.Y. Inverse analysis of residual stress in orthogonal cutting. *J. Manuf. Process* **2019**, *38*, 462–471. [CrossRef]
41. Mirkoohi, E.; Bocchini, P.; Liang, S.Y. Analytical temperature predictive modeling and non-linear optimization in machining. *Int. J. Adv. Manuf. Tech.* **2019**, *102*, 1557–1566. [CrossRef]

crystals

MDPI

Article

The Fracture Behavior and Mechanical Properties of a Support Structure for Additive Manufacturing of Ti-6Al-4V

Sebastian Weber [1,2,*], Joaquin Montero [1,2], Christoph Petroll [2], Tom Schäfer [1], Matthias Bleckmann [2] and Kristin Paetzold [1]

1 Institute for Technical Product Development, University of the Bundeswehr Munich, Werner-Heisenberg-Weg 39, 85579 Neubiberg, Germany; j.montero@unibw.de (J.M.); tom.schaefer@unibw.de (T.S.); kristin.paetzold@unibw.de (K.P.)

2 Bundeswehr Research Institute for Materials, Fuels and Lubricants (WIWeB), Institutsweg 1, 85435 Erding, Germany; christophpetroll@bundeswehr.org (C.P.); matthiasbleckmann@bundeswehr.org (M.B.)

* Correspondence: s.weber@unibw.de; Tel.: +49-89-6004-3685

Received: 1 April 2020; Accepted: 24 April 2020; Published: 27 April 2020

Abstract: In the laser powder bed fusion processes for metal additive manufacturing, a support structure is needed to fix the part to the base plate and to support overhanging regions. Currently the importance of support structure for a successful build process is often underestimated and some effects are not yet well understood. Therefore, this study investigates the fracture behavior and mechanical properties of thin additive manufactured struts using the titanium alloy Ti-6Al-4V and specific machine parameters for support structures. Tensile tests were performed for different strut diameters and the fracture surfaces were analyzed using a laser microscope and a scanning electron microscope. Additionally, the porosity was examined with micro-CT scans. The results were compared with a different set of parameters used for solid parts. The experiments revealed that struts produced with support parameters had no significantly lower tensile strength than the comparative parts. Despite that, some porosity and around two percent of defects on the fracture surface for parts using the solid parameter set have been found. Parts with support parameters show no porosity, even though the energy density is around 30% lower compared to the solid parameter set.

Keywords: support structure; additive manufacturing; Powder Bed Fusion; titanium alloy; Ti-6Al-4V; fracture behavior; mechanical properties

1. Introduction

Additive manufacturing (AM) is a process for the rapid creation of complex parts in small batches. It is associated with a small number of design restrictions and near net-shaped products. Metal additive manufacturing is a manufacturing process with which to create a part based on 3D model data by subsequently adding material layer upon layer [1]. This is in contradiction to subtractive manufacturing methods, which create the final product through material removal usually from a semi-finished part. For machines used in the consumer sector, *3D Printing* is typically used as a synonym to AM. Nowadays its applications range from *rapid prototyping (RP)*, *rapid tooling (RT)* and *rapid manufacturing (RM)* to fully functional parts and even simple assemblies [2]. Based on the ability of AM to create complex shapes, it is usually related to the idea of *design freedom*. Nevertheless some restrictions depending on the used material and process apply [3]. The most common AM technology for industrial metal parts and therefore the focus of this article is laser powder bed fusion (L-PBF). L-PBF uses a laser as an energy source. L-PBF is described as an "AM process in which thermal energy selectively fuses regions of a powder bed" [1].

In L-PBF powder particles of a powder bed are selectively fused together by the laser beam, merging them to a single layer of a part. The beam is guided in the horizontal plane based on trajectories defined through the slicing of a 3D model, which is the method for dividing a 3D model into a finite number of equally spaced layers. After each layer is finished, the build plate is lowered by a fixed amount specifying the layer height. New powder is then placed on top of the previous layer by a recoating device. Repeating this procedure, the part is build layer upon layer. To avoid oxidation, the build chamber is evacuated of remaining oxygen by a constant flow of an inert gas; for instance, argon.

Due to high melting temperatures of the titanium alloy and the extremely high cool down rates, the material tends to expand and contract, leading to high residual stresses. Additionally, during solidification there is not enough time for the microstructure to solidify in an equilibrium phase and thereby resolve the stressed state. These internal stresses lead to severe distortions and deformation of the part [4]. To prevent build failures based on large deformations, a support structure is added for even heat dissipation and to anchor the part on the build plate [5]. After the AM process is finished, these support structures need to be mechanically removed in a post-processing step.

A support structure has different tasks to fulfill in metal additive manufacturing. On the one hand, it is needed to create an evenly distributed heat gradient from the part to the base plate, and on the other hand, deformations need to be reduced or prevented by anchoring the part to the base plate. Another important role is the support of overhanging structures [6], which would otherwise sink down into the loose powder bed. There are different types of support structures used in metal additive manufacturing. Among the most common are block-support, point-support, web-support, contour-support and line-support [6,7]. These types can be implemented in a simple manner with state of the art pre-processing software and are sometimes even generated automatically. More complex approaches use the stacking of unit cells, which is then called a lattice support structure [8,9]. Lately investigations were made with heat and topology optimization, resulting in promising tree-like structures consisting of stems with multiple branches each [10,11]. The benefits of those support structures are the more flexible design and lesser powder consumption, and therefore, build time is reduced when compared to traditional structures. Therefore tree-like support is part of this ongoing research. For this study, the focus is set exclusively on the stem and hence on test samples consisting of multiple struts.

A general rule for a support structure is to use only as much as necessary and as little as possible, due to the fact that an increase in support volume leads directly to longer print times and a higher powder consumption. On the other hand, insufficient support leads to failed print jobs. Knowing the fracture properties of supports and estimating residual stresses in an early stage are key elements for primary support structure design; i.e., for a support that is added during the part design. Due to the current underestimation of the importance and the design of the support structure, the process can fail as a result of severe warping and/or collapse of the part or support. Typically, a support structure is built using manufacturer-provided parameters differing from those for structural parts. Consequently, a thorough investigation of the mechanical and fracture behavior is needed for a proper design. The resulting mechanical properties of the support structures are also very important for structural simulations that lead to new support designs.

Previous research investigated the mechanical properties and fracture behaviors of additive manufactured titanium samples with geometry according to standards [12,13] or other large cross-sectional areas [14], often together with examination of the microstructure [15,16]. Therefore the influences of a small diameter are not taken into account. On the other hand, a lot of effort is put into the design of support structures. Their optimal geometry is analyzed using structural or thermal simulations [7,11], or parameter optimizations are performed [6].

This study aims to fill the gap between support structure design and investigations of mechanical properties for finished parts, to further understand the in-process failure of supports. Therefore, tensile test samples with thin struts, representing the supports, were additively manufactured using the titanium alloy Ti-6Al-4V. Specific machine parameters for support structure were used. Tensile tests were carried out for different strut diameters and build parameters. The resulting fracture surfaces were examined with a scanning electron microscope. To distinguish between pores and trapped particles, additional micro-CT scans were performed. All results were compared with a set of parameters for structural solid parts.

2. Materials and Methods

2.1. Titanium Alloy Powder—Ti-6Al-4V

All test samples used in this study were produced using the titanium alloy powder *Ti64ELI-A LMF* from the manufacturer Trumpf, where ELI (*extra low interstitials*) stands for improved ductility and better fracture toughness. The titanium alloy is also known under the designation Ti-6Al-4V and is one of the most common materials used in metal additive manufacturing [6]. It was produced using plasma atomization, resulting in particle sizes between 15 µm and 45 µm. The particle size distribution is $D10 = 20.6\,\mu m$, $D50 = 33.5\,\mu m$ and $D90 = 43.0\,\mu m$ [17]. The chemical composition of the powder is shown in Table 1.

Table 1. Chemical powder composition [17].

	Al	V	Fe	O	C	N	H	Ti
Min.	5.50	3.50	≤0.25	≤0.13	≤0.08	≤0.05	≤0.012	residual
Max.	6.50	4.50						
Measured	6.49	3.99	0.19	0.09	0.01	0.02	0.0025	balance

Mass fraction in %.

2.2. Test Samples

The test samples were produced with a Trumpf TruPrint 1000 Multilaser L-PBF machine. Two different sets of process parameters were used. The first set's contents are referred to as solid parameters and were used for structural parts. Samples produced with this set are from here on called *solid samples*. The second set is the Trumpf standard for support structure and therefore referred as support (structure) parameters. Samples using this set are henceforth called *support samples*. Both parameter sets use a layer height of 20 µm, a hatch spacing of 80 µm and a laser spot size of 55 µm. The difference between the sets is the used energy input resulting from a laser power of 150 W with a scanning speed of 1600 mm s^{-1} as support structure parameters and 600 mm s^{-1} at 83 W for the solid samples respectively. According to Figure 1, only the solid parameters result in a fully dense part.

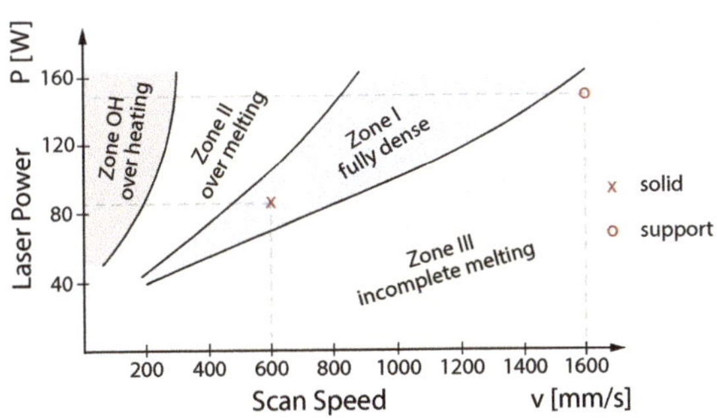

Figure 1. Process parameter relationships for L-PBF manufactured Ti-6Al-4V. Adapted from [18].

The samples had to be designed with a similar outline as the flat version of the ASTM E8 standard tensile test sample for metallic materials with little modifications in the dimensions to fit the clamping mechanism of the tensile machine. The middle section was replaced by the support structure that was to be tested in the tensile test. In this way, the upper and lower block of the sample acted as an aid for the clamping of the sample. A sketch of the geometry is shown in Figure 2. Bar support was used because of its similarity to the stems of tree-like support. For each set of parameters, test samples were produced in an upright orientation with a diameter D range of the struts of 0.1 mm to 1 mm in 0.1 mm steps. The length A was set to 10 mm. The distance between the struts L is the resulting value, when using four struts symmetrically with specified diameters. Samples with a diameter beneath 0.6 mm got heavily distorted during the printing process and were therefore not considered in this investigation. Samples with diameters above 0.7 mm were designed to have only three struts for consistent clearance between the struts and were therefore also excluded from this particular contribution. This leads to a remaining number of 20 test samples, five samples for each diameter and machine parameter configuration. The samples were labeled according to their sizes and diameters as *L10D06* and *L10D07*. With respect to support structure parameters, the letter *S* is added to the end of name for support samples.

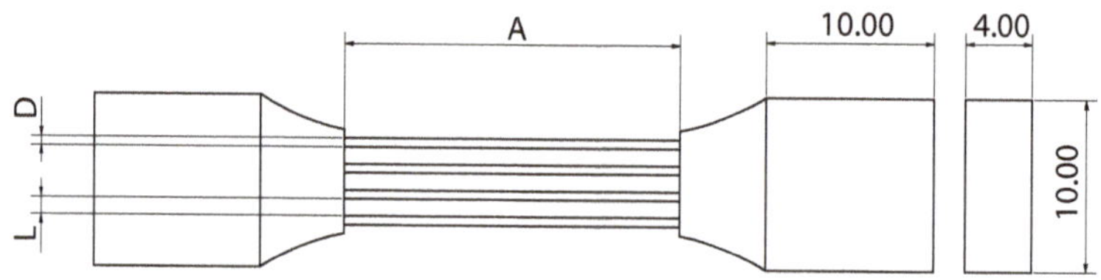

Figure 2. Sketch of the sample geometry.

2.3. Microscopic Analyses

The samples were analyzed using a Keyence 3D laser scanning confocal microscope of the model VK-X200. It is equipped with a violet laser with a wave length of 408 nm. A 10× magnification objective lens was used for the measurements. The captured data have been further processed using a manufacturer provided software. The number of and sizes of defects were analyzed, as were the outer diameter and the diameter of an inner section that was present due to non-fully-molten particles. Additionally the depth information was used to create a 3D model of the fracture surface; exemplary results of this are shown in Figure 3 of Section 3.3.

After the analysis with the confocal laser microscope, the fracture surface of one sample for each parameter set was investigated using a Zeiss Ultra Plus *field emission scanning electron microscope (FE-SEM)*, now referred to as SEM. The acceleration voltage range was 0.02 kV to 30 kV and was set to 1 kV and 1.5 kV for this contribution. A total magnification of 1,000,000× and thereby a resolution of 0.8 nm can be reached with this SEM. As the test samples and the desired images were rather large, the maximum magnification for this investigation was 3000×.

2.4. Micro Tensile Testing

Tensile tests were performed on each sample of the sets in as-built condition without any mechanical post-processing other than removing the samples from the base plate. The tensile testing machine used was a tension/compression module developed and distributed by Kammrath and Weiss GmbH (Dortmund, Germany). The measurement range of the load cell is between 10 µN and 10 kN. The machine uses threaded bars to move one part of the sample fixation with a resolution of 100 nm and speeds of 0.1 µm s^{-1} to 20 µm s^{-1}. For this investigation a constant speed of 1 mm min^{-1} was used. Due to the unpredictable region of the failure on the strut and no available extensometer, the local strain of the struts was not measured.

3. Results

3.1. Geometric Accuracy of the Samples

The diameters of the struts of the test samples were measured before and after the tensile testing. Table 2 shows the measured and the designed values for the diameter of the struts. The standard deviation (σ_d) is given in parentheses. *Side view* is the diameter, based on the thickness of the struts, measured under the confocal laser microscope before the tensile test. Both the *fracture surface* and *core* value were measured after the test from the top view on the fracture surface with the *core* as the fully-molten region, which can be observed later in Figure 3. It can be recognized that the designed diameter is a little bit smaller than the actual one. After the test, the outer diameter of the fracture surface was smaller than the side view value, but still in the range of the designed diameter. Even though the difference between those diameters is in the order of the standard deviation, the reduction might result from necking due to a minor ductile behavior. Lastly the diameter of the fully dense inner section of the fracture surface was determined. This core is around the size of two powder particles smaller than the outside diameter, which results from particles being sintered or partially molten to the outside geometry due to lower heat, where laser paths no longer overlap. This is consistent with the results from the fracture surface analysis in Section 3.3, using the laser and scanning electron microscope.

Table 2. Mean values of the measured diameters (*d*) of the test samples in µm.

Sample	Designed Diameter	Side View	Fracture Surface	Core
L10D06	600	606 ($\sigma_d = 12$)	599 ($\sigma_d = 13$)	480 ($\sigma_d = 13$)
L10D06S	600	627 ($\sigma_d = 7$)	607 ($\sigma_d = 6$)	464 ($\sigma_d = 12$)
L10D07	700	727 ($\sigma_d = 16$)	714 ($\sigma_d = 8$)	565 ($\sigma_d = 13$)
L10D07S	700	726 ($\sigma_d = 19$)	701 ($\sigma_d = 11$)	560 ($\sigma_d = 14$)

3.2. Tensile Properties

The ultimate tensile strength (UTS) was evaluated using the different diameters from Table 2 from before and after the tensile test to calculate the cross-sectional area and the measured forces from the micro tensile test for each sample in as-built condition. UTS_{side} uses the diameter measured from side view, $UTS_{fracture}$ and UTS_{core} respectively the outer and inner diameter measured from the top view of the fracture surface. The results are listed in Table 3. Significantly lower UTS values for the sample L10D06S with support parameters can be seen at first sight while the standard deviation is in the order of ten higher than the others. This results from a failure of some of the test samples at low stress levels without distinct plastic deformation leading to a larger standard deviation as with the other samples. Comparing the L10D07S sample with those for solid parameters no significant deviation can be observed, as the values of solid samples are only 2 % to 4 % higher.

Table 3. Mean ultimate tensile strength (UTS) of the as-built test samples in MPa.

Sample	UTS_{side}	$UTS_{fracture}$	UTS_{core}
L10D06	852 ($\sigma = 11$)	872 ($\sigma = 12$)	1357 ($\sigma = 18$)
L10D06S	640 ($\sigma = 113$)	685 ($\sigma = 119$)	1114 ($\sigma = 245$)
L10D07	839 ($\sigma = 12$)	870 ($\sigma = 13$)	1389 ($\sigma = 20$)
L10D07S	808 ($\sigma = 13$)	867 ($\sigma = 14$)	1358 ($\sigma = 22$)

3.3. Fracture Surface Analysis

In Figure 3 the fracture surface images of the samples L10D07S with and L10D07 without support parameters are shown together with a 3D model of the fracture captured with a confocal laser microscope. In Figure 3a 14 defects can be observed on the fracture surface of the test sample, which was built using the solid parameter set. These defects seem to be pores and are investigated later in Section 3.4. The total area of the pores is around 2 % with respect to the cross-sectional area. The diameters of the pores range from 7.9 µm to 21.3 µm with a mean diameter of 12.9 µm and a standard deviation of 3.5 µm. All of the eight tested struts show the same number of pores. In comparison to (a), Figure 3d shows the same strut of a test sample that was produced with the support parameter set. There is no visible porosity on the fracture surface of Figure 3d. Figure 3b shows the measurement of the outside diameter of the strut from the solid sample, which is listed for all measurements of all samples in Table 2 as *fracture surface*. Figure 3e shows the same measurement for the support sample. In both pictures an outer region with powder particles, which are not fully molten, and an inner circle with the fully dense part, the core, can be observed. The magnitude of the differences between those two diameters is in the order of two powder particles. In Figure 3c,f the 3D model of the fracture is shown. It can be observed that the fracture is angled at almost 45° with a slightly steeper angle in Figure 3f. This angle may result from small variations in the fixation of the samples in the tensile test machine and therefrom result in shear stresses. All in all, it still shows a predominantly brittle behavior. Especially in Figure 3f, some indications of a cup-and-cone fracture mode behavior are visible, leading to the assumption that ductility was present to some extent. The small diameters of the test samples may affect the formation of the cup-and-cone mode. The analysis of brittle and ductile behavior of the test samples will be addressed later in this article.

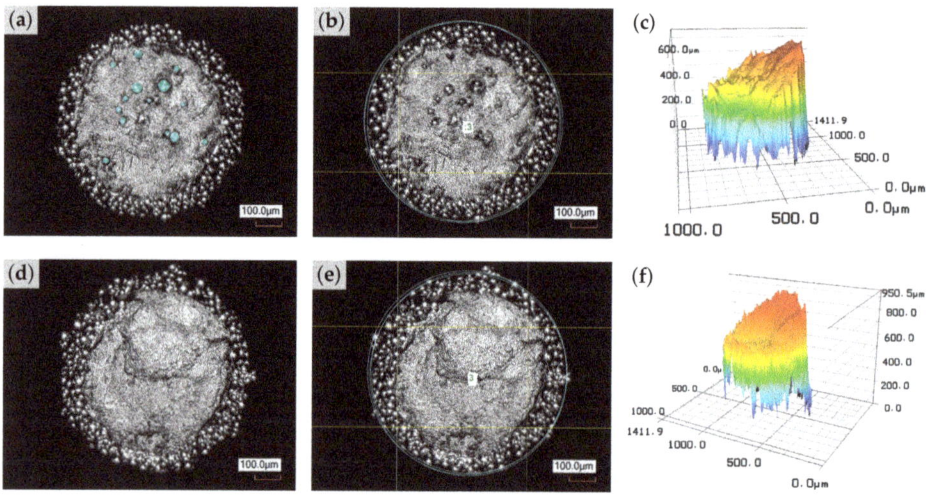

Figure 3. Confocal laser microscopy of the fracture surfaces of the test samples L10D07 (**a**–**c**) and L10D07S (**d**–**f**). Observed pores marked blue in (**a**), not present in (**d**). Outside diameter measurement of the struts in (**b**,**e**). 3D model of each fracture surface respectively in (**c**,**f**).

In Figure 4 different pictures of the fracture surface from the L10D06S sample with support parameters captured by a scanning electron microscope are displayed. At first sight, the hard edges on the fracture surface in Figure 4a indicate a brittle behavior of the fracture. Beside those edges, a single round pore can be observed in the upper half of Figure 4a. The spheres on the outer diameter of the sample, visible in Figure 4a,b, are residual powder particles, which are either semi-molten or sintered to the outside surface. Especially in Figure 4b, it can be observed that the sizes of the particles are mostly consistent, although a few smaller particles are attached to the larger ones. A closer examination of Figure 4b unravels two different types of fracture surface region, one being a brittle section with sharp edges and an obvious uneven surface with differences in height. The other region is more flat and characterized by a micro-dimpled surface that suggests a minor ductile behavior. This can be confirmed with Figure 4c with 3000× magnification. Here the micro-dimpled surface seems again to be very flat with small quasi-cleavage regions with a width of around 2 µm in between the elevated grain boundaries. Bassoli and Denti refer to a similar flat micro-dimpled surface as *quasi-flat rupture morphology* [19]. This micro-dimpled surface combined with tiny quasi-cleavage regions suggests a small granularity of the microstructure matrix. Again, a mixture of brittle and ductile behavior can be observed in Figure 4c with the brittle region with harsh edges in the upper half of the picture and the ductile section, represented by the micro-dimpled surface, in the lower half. Several brittle sections are also visible in Figure 4d. The path of the edges suggests a crack initiation from a ductile outer region towards the center of the strut.

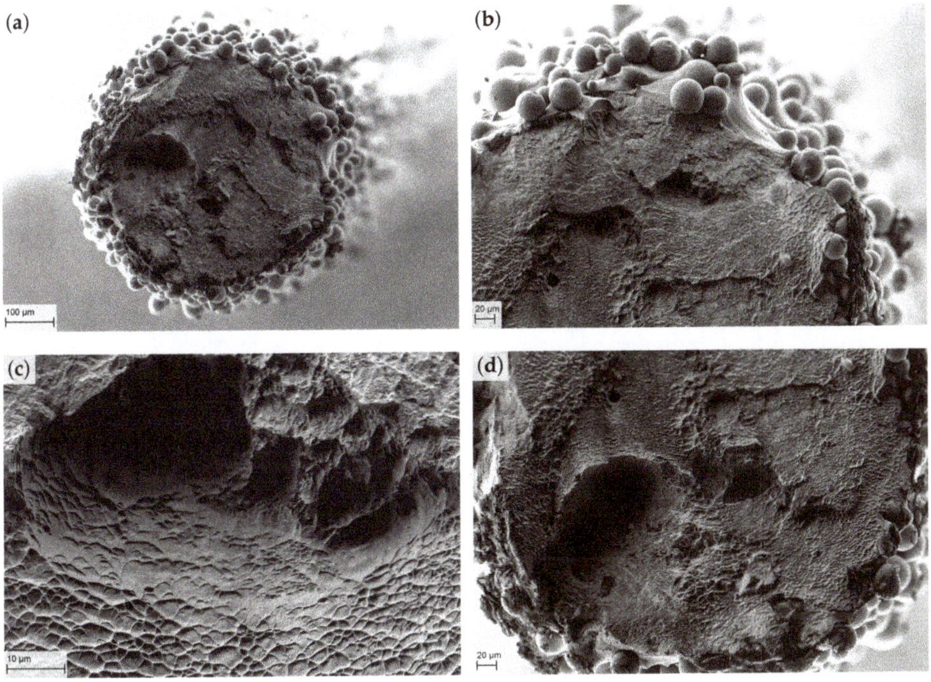

Figure 4. SEM analysis of the support sample L10D06S. (**a**) Fracture surface with 250× magnification. (**b**) Brittle and ductile regions at 500×. (**c**) Micro-dimpled surface at 3000×. (**d**) Mixed fracture at 500×.

The fracture surface of the solid sample L10D06 captured with an SEM is shown in Figure 5. At first sight, there are way more defects or pores visible in Figure 5a compared to Figure 4a. Additionally, a large brittle section can be observed in Figure 5a near to the center of the strut. This brittle region is associated with the number of pores in the fracture layer. For the solid sample, there are again only a few semi-molten and sintered particles attached to the outside surface of the sample, visible in (a,b,d). This shows no difference to the support sample. Brittle regions seem to evolve around pores, as can be observed in Figure 5b, even though it shows a mixed behavior close to the outside surface. Additionally, close to the outer diameter is the very flat micro-dimpled surface of Figure 5c. Around the pore a large quasi-cleavage area can be observed. Figure 5d exhibits a ductile behavior in the outer region in form of a micro-dimpled surface, leading to the assumption of a small amount of necking.

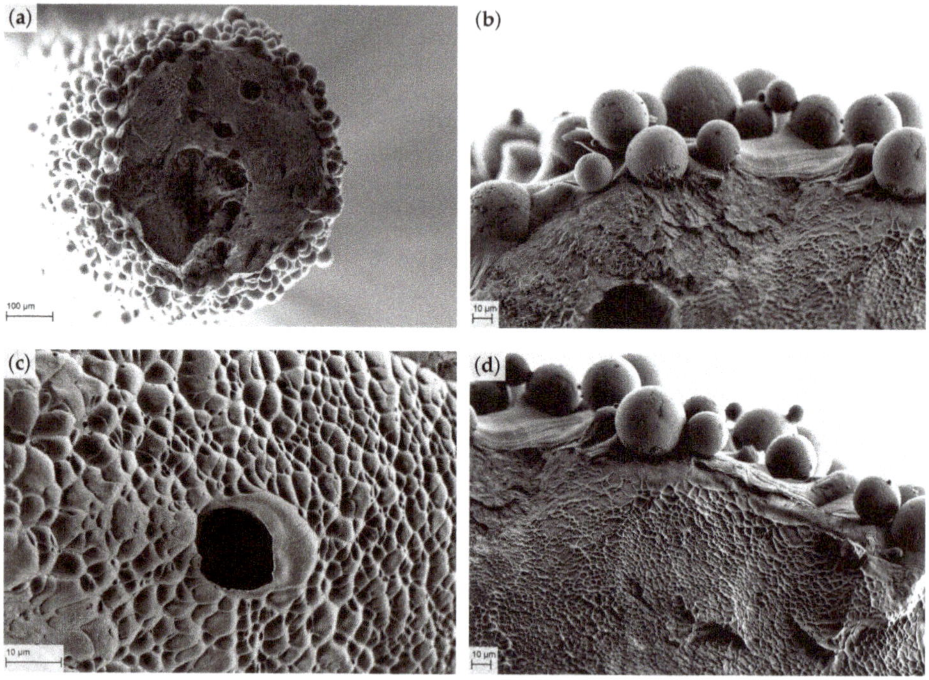

Figure 5. SEM fracture analysis of the solid sample L10D06. (**a**) Overview with pores at 250× magnification. (**b**) Mixed fracture behavior at 1000×. (**c**) Quasi-flat dimpled surface with pore at 3000×. (**d**) Dimpled surface close to the outside surface at 1000×

During the Investigations of the fracture surface, martensite columns could be observed in regions of semi-molten particles on the outside surface. Figure 6a shows the location to be close to the edge of the fracture surface with a semi-molten particle attached to the outside surface. In Figure 6b the actual martensite columns with a length of around 2 µm and an approximate width of 50 nm can be observed in a 90° pattern on the upper half of the picture and a brittle section of the fracture surface in the lower half.

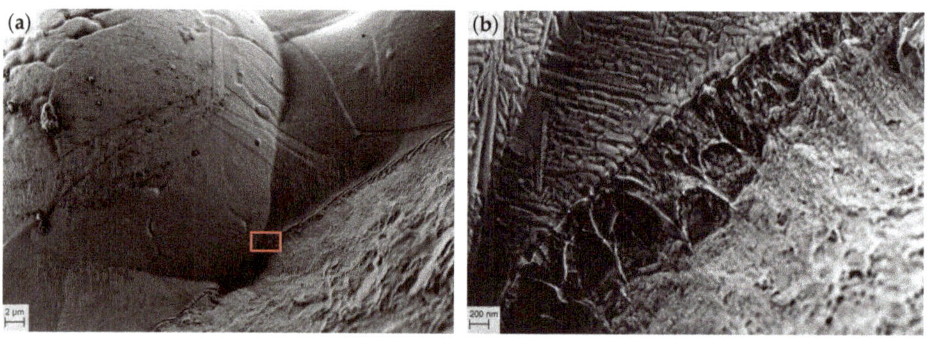

Figure 6. SEM analysis of the boundary surface of the test sample L10D06. (**a**) Semi-molten powder particle. (**b**) Magnification of (**a**) showing martensite columns on non fractured surface.

3.4. Porosity

In order to distinguish between pores and trapped particles that fell out after the tensile test, a micro-CT scan of both samples was performed. Figure 7a shows the scan image of one half of the solid sample L10D07 through the center of the specimen. Figure 7b displays the right two struts close to the rupture. Several small pores with diameters between 10 µm and 40 µm can be observed. Scans of the support sample showed, as expected, no signs of pores, and they were therefore not included in this article.

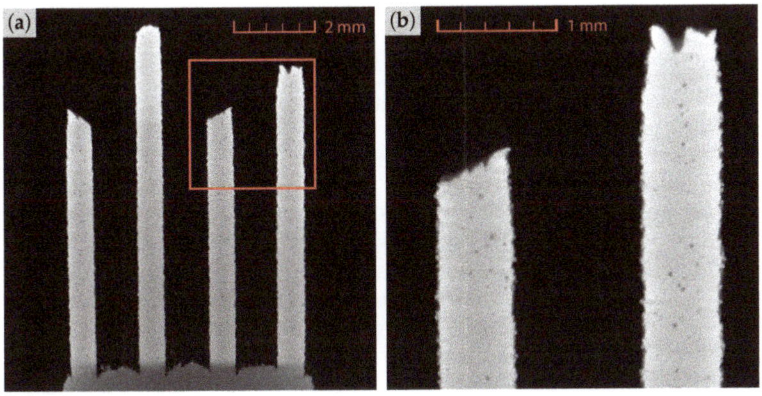

Figure 7. Micro-CT scan of the solid sample L10D07 with porosity. (**a**) Overview of all four struts. (**b**) Magnification with visible pores

Table 4 lists the mean properties of the porosity of the samples with 0.6 mm and 0.7 mm diameter struts for both parameter sets. It is clearly visible that samples produced with the solid parameter set contain more pores and thereby have larger defect surfaces. In total there is around 2 % of the fracture surface subject to porosity. There is no obvious reason for the solid parameter combination to show porosity, as the combination is located in the fully dense region of Figure 1. Still, it might result from an already too high energy input or the fact that small amounts of porosity were not considered in the definition of the fully dense zone. Occasional pores in the support samples seem to have no significant differences in diameter. The sample L10D07S showed a single pore in one of the four struts. As a mean value, this leads to an amount of zero pores, but a remaining defect surface of 61 µm^2 still.

Table 4. Mean properties of pores, visible on the fracture surfaces of the test samples.

Sample	Number of Defects	Defect Surface [µm^2]	Percent of Total Area	Pore Diameter [µm]
L10D06	10	6267 ($\sigma_s = 2075$)	2.23 %	15.1 ($\sigma_d = 2.9$)
L10D06S	2	769 ($\sigma_s = 2917$)	0.36 %	11.4 ($\sigma_d = 3.3$)
L10D07	12	6970 ($\sigma_s = 2924$)	1.74 %	12.3 ($\sigma_d = 3.7$)
L10D07S	0	61 ($\sigma_s = 106$)	0.02 %	-

4. Discussion and Conclusions

In this article, the characteristics of thin struts, which act as a support structure, were analyzed for two sets of build parameters. Tensile tests and subsequent fracture surface analysis were performed. It was shown that semi-molten and sintered particles on the outside surfaces of the samples affect the load bearing of the parts as the designed dimensions include these regions with particularly lower bonding strength. Small amounts of porosity in the parts showed no effect on the overall tensile strength.

Compared to round tensile test specimens produced according to the ASTM standards, the ultimate tensile strength evaluated in this article, is significantly lower. For cross-sectional areas calculated with the outside diameters of the specimens, the UTS was below 900 MPa and thereby strongly deviated from the range reported by Kasperovich and Hausmann given as 1040 MPa to 1062 MPa [20]. On the other hand, if only the core area is considered, values above 1300 MPa are even higher than the strength of as-built and machined samples, for which Vilaro et al. evaluated 1166 MPa [21] and Rafi et al. 1219 MPa [12]. This confirms that the outside, non-fully-molten region carries significantly less load than the core region, while still contributing to the overall load bearing. The values for L10D06S are considered statistical outliers, as the failure at low stress levels might result from a slightly tilted clamping or minor defects in the sample. The high standard deviation indicates that it was an unusual failure and if added to the UTS value it results in the same range as the other values.

The fracture surface analysis revealed a mixed fracture behavior showing micro-dimpled regions for ductile and quasi-cleavage sections for brittle failure. This is in accordance with results from Krakhmalev et al. [13] and Bassoli and Denti [19]. The cup-and-cone shape of the necking region reported by Krakhmalev et al. [13], however, was only present to some extent for the struts with small diameters. Additionally no distinct necking could be observed, as the diameter reduction was only in the range of the standard deviation of the measurements.

The results discussed in this article helped us to further understand the behavior of a support structure built with specific support parameters compared to solid parameters. Using these results, support structures can be investigated in process simulations with adapted material characteristics or be designed to meet load bearing needs. In addition, it was shown that the small diameters of the struts affected the fracture behaviors and tensile strengths of the samples. The authors suggest for future research to investigate the effect of the diameter on the micro structure, especially the differences between outside regions and the cores of the samples.

Author Contributions: Conceptualization, S.W. and J.M.; data curation, S.W. and T.S.; investigation, S.W., C.P. and T.S.; resources, C.P.; supervision, K.P.; writing—original draft, S.W. and J.M.; writing—review and editing, M.B. and K.P. All authors have read and agreed to the published version of the manuscript.

Funding: This research received no external funding.

Acknowledgments: The authors like to thank the 3D Printing Center at the Bundeswehr Research Institute for Materials, Fuels and Lubricants (WIWeB) for their assistance with the L-PBF print of the test samples and for providing the CT and SEM images.

Conflicts of Interest: The authors declare no conflict of interest.

References

1. ISO. *ISO/ASTM52900-15, Terminology for Additive Manufacturing-General Principles-Terminology*; Technical Report; ASTM International: West Conshohocken, PA, USA, 2015. [CrossRef]
2. 3D Printing Parts for the Aviation and Aerospace Industries. Available online: https://www.ge.com/additive/additive-manufacturing/industries/aviation-aerospace (accessed on 13 February 2020).
3. Ponche, R.; Kerbrat, O.; Mognol, P.; Hascoet, J.Y. A novel methodology of design for Additive Manufacturing applied to Additive Laser Manufacturing process. *Robot. Comput.-Integr. Manuf.* **2014**, *30*, 389–398. [CrossRef]
4. Mercelis, P.; Kruth, J. Residual stresses in selective laser sintering and selective laser melting. *Rapid Prototyp. J.* **2006**, *12*, 254–265. [CrossRef]
5. Cloots, M.; Spierings, A.; Wegener, K. Assessing new support minimizing strategies for the additive manufacturing technology SLM. In Proceedings of the Solid Freeform Fabrication Symposium (SFF), Austin, TX, USA, 12–14 August 2013; pp. 12–14.
6. Calignano, F. Design optimization of supports for overhanging structures in aluminum and titanium alloys by selective laser melting. *Mater. Des.* **2014**, *64*, 203–213. [CrossRef]

7. Krol, T.A.; Zaeh, M.F.; Seidel, C. Optimization of Supports in Metal-Based Additive Manufacturing by means of Finite Element Models. In Proceedings of the SFF Symposium Preceedings, Austin, TX, USA, 6–8 August 2012; pp. 707–718.
8. Strano, G.; Hao, L.; Everson, R.M.; Evans, K.E. A new approach to the design and optimisation of support structures in additive manufacturing. *Int. J. Adv. Manuf. Technol.* **2013**, *66*, 1247–1254. [CrossRef]
9. Hussein, A.; Hao, L.; Yan, C.; Everson, R.; Young, P. Advanced lattice support structures for metal additive manufacturing. *J. Mater. Process. Technol.* **2013**, *213*, 1019–1026. [CrossRef]
10. Malekipour, E.; Tovar, A.; El-Mounayri, H. Heat Conduction and Geometry Topology Optimization of Support Structure in Laser-Based Additive Manufacturing. In *Mechanics of Additive and Advanced Manufacturing, Volume 9*; Wang, J., Antoun, B., Brown, E., Chen, W., Chasiotis, I., Huskins-Retzlaff, E., Kramer, S., Thakre, P.R., Eds.; Springer International Publishing: Cham, Switzerland, 2018; pp. 17–27. [CrossRef]
11. Bartsch, K.; Herzog, D.; Emmelmann, C.; Lange, F. A Novel Approach to Support Structures Optimized for Heat Dissipation in SLM by Combining Process Simulation with Topology Optimization. In Proceedings of the NAFEMS World Congress 2019, Quebec City, QC, Canada, 17–20 June 2019; p. 25.
12. Rafi, H.K.; Karthik, N.V.; Gong, H.; Starr, T.L.; Stucker, B.E. Microstructures and Mechanical Properties of Ti6Al4V Parts Fabricated by Selective Laser Melting and Electron Beam Melting. *J. Mater. Eng. Perform.* **2013**, *22*, 3872–3883. [CrossRef]
13. Krakhmalev, P.; Fredriksson, G.; Yadroitsava, I.; Kazantseva, N.; Plessis, A.D.; Yadroitsev, I. Deformation Behavior and Microstructure of Ti6Al4V Manufactured by SLM. *Phys. Procedia* **2016**, *83*, 778–788. [CrossRef]
14. Simonelli, M.; Tse, Y.Y.; Tuck, C. Effect of the build orientation on the mechanical properties and fracture modes of SLM Ti–6Al–4V. *Mater. Sci. Eng. A* **2014**, *616*, 1–11. doi:10.1016/j.msea.2014.07.086. [CrossRef]
15. Popovich, A.; Sufiiarov, V.; Borisov, E.; Polozov, I. Microstructure and Mechanical Properties of Ti-6Al-4V Manufactured by SLM. *Key Eng. Mater.* **2015**, *651-653*, 677–682. [CrossRef]
16. Yadroitsev, I.; Krakhmalev, P.; Yadroitsava, I.; Du Plessis, A. Qualification of Ti6Al4V ELI Alloy Produced by Laser Powder Bed Fusion for Biomedical Applications. *JOM* **2018**, *70*, 372–377. [CrossRef]
17. Trumpf. *Batch Test Report—Ti64Eli—A 17-E361*; Trumpf GmbH + Co. KG: Ditzingen, Germany, 2019.
18. Gong, H.; Rafi, K.; Gu, H.; Starr, T.; Stucker, B. Analysis of defect generation in Ti–6Al–4V parts made using powder bed fusion additive manufacturing processes. *Addit. Manuf.* **2014**, *1–4*, 87–98. [CrossRef]
19. Bassoli, E.; Denti, L. Assay of Secondary Anisotropy in Additively Manufactured Alloys for Dental Applications. *Materials* **2018**, *11*, 1831. [CrossRef] [PubMed]
20. Kasperovich, G.; Hausmann, J. Improvement of fatigue resistance and ductility of TiAl6V4 processed by selective laser melting. *J. Mater. Process. Technol.* **2015**, *220*, 202–214. [CrossRef]
21. Vilaro, T.; Colin, C.; Bartout, J.D. As-Fabricated and Heat-Treated Microstructures of the Ti-6Al-4V Alloy Processed by Selective Laser Melting. *Metall. Mater. Trans. A* **2011**, *42*, 3190–3199. [CrossRef]

crystals

Review

In Situ Monitoring Systems of The SLM Process: On the Need to Develop Machine Learning Models for Data Processing

Pinku Yadav [1,2], Olivier Rigo [1], Corinne Arvieu [2], Emilie Le Guen [2] and Eric Lacoste [2,*]

1 SIRRIS, Rue du Bois Saint-Jean 12, 4102 Seraing, Belgium
2 CNRS, Arts et Metiers Institute of Technology, University of Bordeaux, Bordeaux INP, INRAE, I2M Bordeaux, F-33400 Talence, France
* Correspondence: Eric.lacoste@u-bordeaux.fr; Tel.: +33-05-5684-5865

Received: 27 May 2020; Accepted: 13 June 2020; Published: 18 June 2020

Abstract: In recent years, technological advancements have led to the industrialization of the laser powder bed fusion process. Despite all of the advancements, quality assurance, reliability, and lack of repeatability of the laser powder bed fusion process still hinder risk-averse industries from adopting it wholeheartedly. The process-induced defects or drifts can have a detrimental effect on the quality of the final part, which could lead to catastrophic failure of the finished part. It led to the development of in situ monitoring systems to effectively monitor the process signatures during printing. Nevertheless, post-processing of the in situ data and defect detection in an automated fashion are major challenges. Nowadays, many studies have been focused on incorporating machine learning approaches to solve this problem and develop a feedback control loop system to monitor the process in real-time. In our study, we review the types of process defects that can be monitored via process signatures captured by in situ sensing devices and recent advancements in the field of data analytics for easy and automated defect detection. We also discuss the working principles of the most common in situ sensing sensors to have a better understanding of the process. Commercially available in situ monitoring devices on laser powder bed fusion systems are also reviewed. This review is inspired by the work of Grasso and Colosimo, which presented an overall review of powder bed fusion technology.

Keywords: L-PBF; in situ sensing; quality assurance; machine learning

1. Introduction

Additive manufacturing (AM), colloquially referred to as 3D printing, is defined by ASTM standard ISO/ASTM 52900 as "process of joining materials to make parts from the 3D model data, usually layer upon layer, as opposed to subtractive manufacturing and formative manufacturing methodologies" [1]. In recent years, the manufacturing industry has seen a paradigm shift due to the full range of advantages of AM, such as complexity-free design, no need for tooling, reduction in waste, and amount of production steps compared to conventional techniques. Fabrication of complex geometries without increasing the production cost is one of the capabilities of the AM. This capability of AM allows the designing of complex lightweight components without compromising the functionality. For example, in the aerospace and automotive industries, the reduction in material weight is directly linked to the reduction in fuel consumption. Therefore, the AM has matured very rapidly in terms of technical aspects and process aspects. These developments have led to convincing many industries to adopt AM for manufacturing novel parts across many sectors.

Despite many technological advantages, many problems related in particular to the control of defects (porosity rate, cracking, etc.), to the management and control of heat transfer during the process, or material supply problems persist. Several operating factors, such as processing parameters, part design, support placement, powder properties, etc., can influence the quality of the finished part [2]. For example, Galy et al. [3] reviewed all the factors in the L-PBF process that influence the final quality of Al alloy parts. All these factors hinder the full adaption of the AM by many niche industries [4–6]. Many of the factors that can influence the quality of the final part (Figure 1) can be controlled in-line, and the process signatures can be directly linked to the quality [7–9].

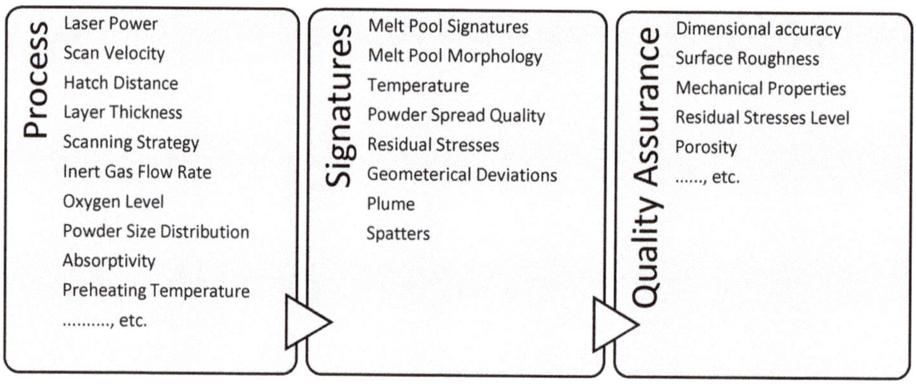

Figure 1. List of process parameters, and corresponding signatures which influence the product quality [2].

This led to the development of in situ monitoring of the process and closed-loop control. There is an enormous number of in situ sensing techniques developed for all types of AM processes. However, we will focus mainly on the laser powder bed fusion process (L-PBF), also known as selective laser melting (SLM). Usually, melt pool monitoring is the most used technique to control the L-PBF process due to the dependency of melt pool dynamics on the quality metric of the part [10].

Generally, we monitor the process at three different levels, which are:

1. Local melt pool monitoring: involves monitoring of the melt pool morphology (shape and size) and heat-affected zone.
2. Layer wise monitoring: The total temperature distribution of the printed layer is monitored. The homogeneous temperature profile of the layer shows the stability of the process.
3. Part volume monitoring: it involves the monitoring of the whole build volume at the end of the process.

The in situ sensors for L-PBF can be divided into acoustic, optical systems, which are similar to the laser welding literature [11]. Acoustic sensors are complicated to calibrate due to the significantly high noise factor. Therefore, optical systems, such as cameras and photodiodes, are common for the L-PBF process. Clijsters et al. [12] and Craeghs et al. [13–15] have developed a co-axial melt pool monitoring system in-house for the L-PBF process in the visible-near infrared detection range. Demir et al. [16] proposed a co-axial system with a set of focused optics to control the area of focus. It is critical to control the area of focus to enhance the resolution around the melt pool. Nowadays, many commercial L-PBF machines are equipped with in situ monitoring systems. Nevertheless, the commercial systems are very closed and act as a black box. Therefore, it is essential to post-process the acquired sensors data to predict the quality of the part.

There are good critical reviews available on in situ monitoring and factors affecting the powder bed fusion (PBF) process [7–9]. Therefore, there is a lack of comprehensive review, especially focused on the L-PBF process and covers the process, in situ monitoring systems, and post-processing of in situ sensing data. This review article bridges the gap and provides a brief overview of the in situ monitoring systems and treatment of data in the L-PBF process. Sections 3 and 4 are inspired by the work of Grasso et al. [9], which discussed commonly occurred defects in the powder bed fusion process. However, we have only reviewed the defects related to the L-PBF process.

This article is organized as follows. Firstly, we discuss the working principle of the L-PBF process and the most common defects that occur during the process, followed by the working principle of the in situ sensing devices. Finally, we review the literature available on the data treatment and automatic detection of defects in the L-PBF process using machine learning.

2. Working Principle of L-PBF

Laser powder bed fusion (L-PBF), also known as selective laser melting (SLM) and Direct melt laser sintering (DMLS), is an additive manufacturing process in which a laser source (thermal energy) selectively fuses regions of powder bed in a layer by layer fashion [17]. The schematic of L-PBF is shown in Figure 2a. The complex thermomechanical process in the form of a melt pool occurs locally due to laser–material interaction [18]. The schematic of the laser–material interaction region is shown in Figure 2b [19]. The non-uniformity of temperature distribution in the melt pool (center of melt pool is at a higher temperature) develops a surface tension gradient, which leads to the thermocapillary motion of the liquid, i.e., the liquid from the center of the melt pool is transported to the edges (colder region). When the laser source moves away from the molten pool, a negative surface tension gradient develops in the melt pool, which generates a shallow and well-distributed liquid mass. Similarly, the segregated melt at the melt pool edges will acquire sufficient surface energy, which then flows back to the hotter region, and the convection loop completes [20]. Rapid solidification with cooling rates of 10^6 K/s of melt pool allows for interlayer bonding [21].

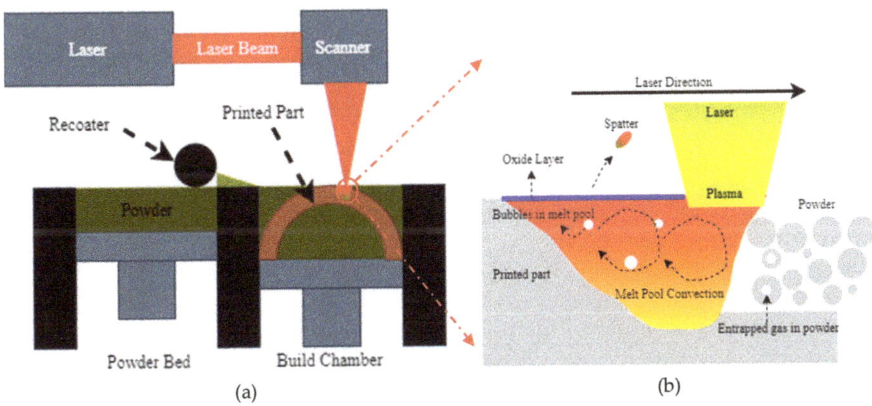

Figure 2. (**a**) Schematic diagram of L-PBF process [17], (**b**) Laser–melt interaction region [19].

For metallic alloys, the support structures are required to anchor parts and their features to the build plate. It is necessary because the thermal gradients in the building part are high, which can lead to thermal stresses and warping if the anchors are not used. The thick build plate serves as the heat sink and also prevents the parts from warping while printing. The preheating of the build plate, which depends on the material manufactured, is done to minimize thermal stresses in the part. The dimensional accuracy and ability to print good surface finish products and fine feature details is one of the advantages of the L-PBF process.

3. Defects

As we discussed, the L-PBF process is a complex thermomechanical process that is affected by many controllable process parameters such as laser power, scan speed, hatch distance, powder material, and powder morphology. All these parameters can strongly influence the laser–material interaction, which will influence the thermophysical mechanism, resulting in various physical phenomena such as material evaporation, microstructural evolution, melt pool instabilities, and thermal stresses. Thus, the formation of various defects during the process is inevitable. Here, we review the formation mechanism of the most prevalent defects, such as porosities, lack of fusion, balling, and cracking defects that can be detrimental to the mechanical properties of the final part.

3.1. Porosities

Spherical and non-spherical porosities, which are usually smaller in size (100 μm), are the most common defects in the L-PBF process, which significantly influence the fatigue properties of the part [22–27]. The leading cause for the formation of gas porosities (generally spherical, as shown in Figure 3a) is the entrapping of gas in the melt pool due to rapid solidification. The rapid solidification of the melt pool does not allow dissolved gas to come out on the surface. As we know, at high temperatures, the gas dissolves in the melt pool very easily. Moreover, enrichment of gas in the liquid melt can be due to various factors such as low packing density of powder bed, gas inclusion during gas atomization of powders, and evaporation due to high laser power. Dissociation of oxide films, adsorbed gas, and moisture on the powder material also results in gas formation [28]. For example, Gong et al. [29] showed the presence of gas porosities due to the entrapping of gas bubbles, which originated due to the vaporization of the low melting material in the alloys. The gas bubbles could not escape to the surface due to rapid solidification. Zang et al. [30] reported the random distribution of the porosities within the printed layer, between the adjacent layers, and on the surface of the part.

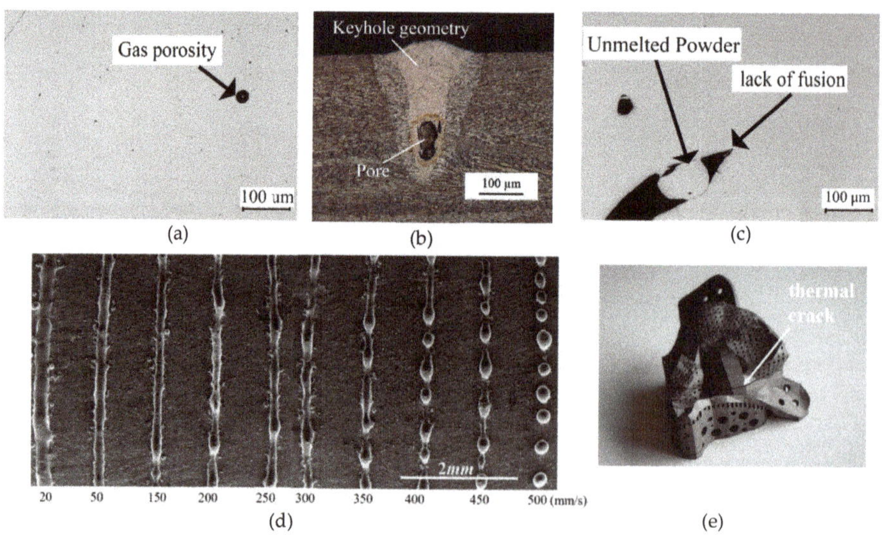

Figure 3. Micrographs of defects observed in L-PBF, such as (**a**) gas porosities [22], (**b**) key hole porosity [29], (**c**) lack of fusion defect [22], (**d**) balling, and (**e**) thermal crack. Reused under Creative Commons Attribution License.

Another type of porosity called "keyhole porosity" is due to excessive energy input, which leads to temperatures beyond the boiling point of the material, which causes evaporation of the material and forms a plasma. The laser beam penetrates the powder bed deep through the vapor voids, which creates a large melt pool. High energy density can occur at lower scan speeds, where the longer exposure time initiates boiling, which unstabilizes the melt pool and induces small metallics balls. The intense recoil pressure creates denudation along the scan track and, together with Marangoni convection (thermocapillary convection), forms a melt pool depression into the powder layers. The breakdown of the melt pool sidewalls during solidification leads to entrapped irregular porosities in the printed layer (as shown in Figure 3b) [31].

Still, there is a lack of universal consensus among researchers on the evolution mechanism of keyhole porosities [29]. For example, according to reference [25], the keyhole porosities occur when the Marangoni effect outweighs the buoyancy. Similarly, Choquet et al. [32] stated that vortices, high fluid speed, and recirculation in the melt pool leads to keyhole phenomena. It is challenging to observe this type of defect experimentally and requires a sophisticated set up to observe it on-line. For instance, Cunningham et al. [33] used ultrahigh-speed X-ray imaging to capture keyhole evolution in titanium alloy while printing. They have explained the evolution of the keyhole porosity, which is as follows: As the laser turned on, the powder began to melt, and a solid–liquid interface was formed. Once the melt pool temperature reached near the boiling point of the material, localized vaporization led to recoil pressure and formed a depression. Due to recoil pressure, the melt pushed up and out of depression, which developed instability in the melt pool. Soon after, the shallow vapor depression transitioned to deep, conical depression. Then, deep vapor depression rapidly penetrated the melt pool, and displacement of the melted liquid from the center of the melt pool occurred. The displacement in the liquid introduced the liquid–vapor interface fluctuations and changed the melt pool shape from quasi-semicircular to a bimodal shape to form keyhole porosity.

3.2. *Lack of Fusion*

Another common defect observed in L-PBF is the lack of fusion defect (LOF), which occurs due to incomplete fusion of two adjacent layers (as shown in Figure 3c). It can happen due to (i) insufficient energy input leading to incomplete melting of the powder layer [34], (ii) insufficient material due to shrinkage during solidification, and (iii) poor bonding between layers due to oxidation, which influences the wetting angle between layers [29,35]. As the L-PBF process selectively melts the powder line by line and layer by layer, the LOF defects are usually present in between the layers or between the scan vectors. Sometimes, the uneven powder layer spreading in consecutive layers also leads to incomplete melting of the powder, which also results in an LOF defect in the final part. Clijsters et al. and Read et al. [12,36] showed that the rapid formation of an oxide layer on the solidified layer of AlSi10Mg leads to poor wettability in between layers, which results in an LOF defect.

3.3. *Balling*

Another common defect in the L-PBF process is called "balling", which leads to disruption in interlayer bonding, poor surface quality, and porosity in the final part [36]. It occurs mainly in high scan speeds, which leads to an elongated melt pool that breaks into small beads (10 μm) to reduce the surface tension due to Rayleigh instability. Additionally, partially melted powder due to low input energy leads to limited melt formation and produces coarser sized beads (500 μm). Oxide layer or presence of moisture on the powder surface can also disrupt the melt pool surface tension, resulting in balling. These impurities reverse the capillary motion at the melt pool surface, and positive surface tension gradient forces the melt pool to flow inwards to the center, which develops unfavorable wetting conditions and leads to spheroidization. Remelting/double exposure or deoxidizing agents can significantly reduce the balling phenomena [37,38]. Li et al. [39] presented the SEM images of the balling phenomena for stainless steel w.r.t different scanning speeds in the L-PBF process (as shown in Figure 3d). Other authors, for example [38,40], reported the balling phenomenon in L-PBF for different materials.

3.4. Cracks

The residual stresses are the stresses that remain in the part after fabrication; these stresses are significantly detrimental to the mechanical strength of the final part. Higher thermal gradient due to rapid heating and cooling cycles develops residual stresses in the part and results in delamination, part distortion, and warping [41,42]. The two mechanisms responsible for the evolution of residual stresses are temperature gradient mechanism (TGM) and rapid solidification mechanism. During TGM, the pre-solidified layer underneath the melted layers is heated up very rapidly upon laser exposure, which tries to expand but is restricted by the cold solidified material. This restriction in thermal expansion develops compressive residual stresses at the regional heat affected zone (see Figure 4a) [43]. During the solidification mechanism, the melted zone cools very rapidly, and shrinkage is again restricted by the underlying layer, developing tensile residual stresses in the upper layer (Figure 5a). Therefore, the cyclic heating and cooling phases during printing results in larger tensile residual stresses in the top surface layers and residual compression stress at the bottom of the printed part (Figures 4b and 5b) [43].

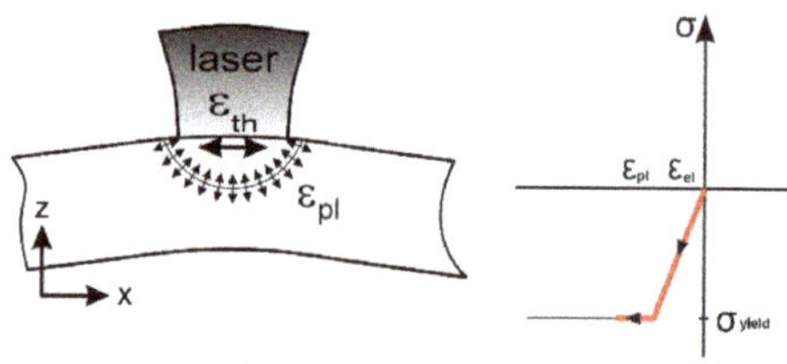

Figure 4. (**a**) Temperature gradient mechanism during laser heating, (**b**) representation of the evolution of thermal stress and strains in the irradiated zone [43].

Figure 5. (**a**) Evolution of stresses during solidification phase, (**b**) representation of the evolution of residual stresses and strain in the irradiated zone [43]. Reused under Creative Commons Attribution License.

At a given time and temperature, when the tensile residual stresses cross the ultimate tensile strength of the material, it leads to cracking in the part [44]. As reported [45], the crack initiation and propagation start on the as-built surface as it adhered to the partially melted material and leads to interlayer delamination, also called interlayer cracking. Materials like nickel-based alloys, copper, and stainless steel are highly susceptible to cracking due to their low thermal conductivity [46–48]. The pre-heating of the build plate is often applied to reduce the effect of the residual stresses in the final part.

3.5. Part Distortions

Another critical aspect of residual stresses on the overall quality of the part health is "part distortion". If the thermally induced local residual stresses exceed the yield stress of the material, it can lead to part distortion in the middle of the process. Part distortion can also damage the recoater permanently. Kruth et al. [49] reported the warping in the SLMed parts due to high thermal fluctuations experienced by the material in a short interval of time. The relaxation in the residual stresses after the removal of the finished part from the base plate and support structures, leads to macroscopic distortion in the final part (as shown in Figure 3e). Mugwagwa et al. [50] studied the influence of process parameters such as laser power, scanning speed, scanning strategies on the evolution of residual stresses, and part distortion experimentally and with finite element modeling. Sometimes, the warping of the parts can lead to failure of the bolts of the build plate as well. Mishurova et al. [51] studied the influence of support structures on part distortion using the X-ray diffraction method. The stress redistribution takes place for the samples with support structures after removal from the base plate, whereas for parts without support structures, only simple stress relaxation observed. They have concluded that the support structures decrease the amount of residual stresses in the part.

Other defects or anomalies such as surface defects, material contamination, and geometrical inaccuracy are not discussed in this article. However, interested readers can refer to references [52–55].

4. Origin of Defects

L-PBF process involves a large number of critical parameters that can influence the quality of the final part and leads to the above mention defects if not optimized. Sharratt et al. [26] categorized the influencing parameters into three categories, which are (i) process defects, (ii) equipment-induced defects, (iii) design defects. Equipment defects are basically due to aberration in the equipment of the machine such as f-theta lens aberration, improper calibration, wear of recoating blade, etc. Nowadays, due to significant technological advancements, most of the commercial machines are well-calibrated and standardized to minimize such defects. Design defects are due to improper optimization of the design as per AM standards such as lack of support structures, overhang angle, and bad orientation. These defects can be minimized by using commercial software such as AdditiveLab, Simufact, Virfac, etc., which helps to find the best support structure placement based on the thermomechanical simulations.

A lot of time and money is consumed to find best-fit process parameters such as laser power, scan speed, hatch distance, layer thickness, etc., to reduce the third type of defects called "process defects". All the parameters mentioned above can be combined in the empirical formula of 'Energy density (E)' given by Equation (1), where P, v, h, and t represent laser power, scan speed, hatch distance, and layer thickness, respectively.

$$E = \frac{P}{v \times h \times t} \tag{1}$$

Each paramete in Equation (1) has its effect on the final quality of the part. For example, at lower scan speed and high laser power, the energy density is higher, and its attributes to porosity defects are due to entrapment of gas. Thijs et al. [27] stated that high energy input in Ti6Al4V powders leads to large melt pool size, which causes denudation. Thus, not enough material is available to fuse the adjacent scan vector, which leads to bigger pores in the part. The higher residual stresses are also reported for low scan speed and higher laser power [46].

Similarly, at low laser power and higher scan speed, a lack of fusion defects is more pronounced due to a lack of energy input to fully melt the powder layer [29,35,56]. Sun et al. [57] showed the relationship between the part density and input energy density for Al0.5CoCrFeNi High Entropy Alloys (Figure 6). It can be observed that for lower energy density, the highest porosity is reported. Therefore, it is crucial to optimize the process parameters to attain the maximum dense parts.

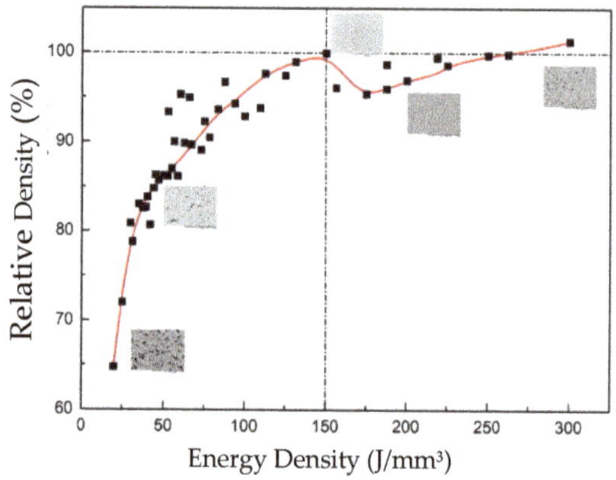

Figure 6. Relative density as a function of energy density [57]. Reused under Creative Commons Attribution License.

It is important to note that the scanning strategy also makes an impact on the overall quality of the part. Although it has not been considered in the energy density equation, the scanning strategy directly influences the heat balance of the part. For example, Aboulkhair et al. [30] showed the relation between the scan speed and different scan strategies, as shown in Figure 7. The different strategies used are: "2X" is similar to "X" but each layer is scanned twice, "X&Y 2HS" means that each layer was scanned twice having each scan pass perpendicular to the one before and with different hatch spacings for each scan, "pre-sinter" means to first scan the layer with half the power followed by a second scan with full power, "overlap" strategy indicates where each layer was scanned twice with the second scan melting the overlap between each two adjacent melt pools, and "alternating" strategy indicates scanning each layer in a direction rotated by 90° to its precursor. As can be concluded from Figure 7, the pre-sinter scanning strategy and scanning speed of 500 mm/s report a high density of AlSi10Mg parts.

5. Melt Pool Signatures

Laser–material interaction in L-PBF is a complex thermophysical phenomenon occurring in a short interval of time. The shape and size of the melt pool are significantly influenced by the process parameters and geometrical factors as well. Therefore, melt pool signatures provide essential information about the quality of the part. A different range of electromagnetic spectrum emanates from the melt pool, which serves as the input variable for in situ sensing devices. The types of electromagnetic radiations observed from the melt pool are shown in Figure 8. As we know, scattering, absorption, and diffraction of the laser light are the common phenomena that occur. Hence, the backscattering of the laser light is a common phenomenon that occurs in L-PBF. Therefore, it is vital to eliminate the backscattering of the laser light in the in situ sensing module by installing correct mirrors. Thermal emissions from the melt pool occur at the wavelengths that correspond to the melt temperature. High laser energy can lead to the evaporation of melt and ionization of the gas, which results in the formation of a plasma plume above the melt pool. The electromagnetic radiation from the plasma can also add noise to the captured signatures [58–61]. Sometimes, the plasma also hinders the laser beam, which results in low input energy to the powder bed. Thermal radiation, which ranges from 900 to 2300 nm, is captured to monitor the process and links it to the temperature field of the melt pool. Plasma radiations, which are in the range of visible wavelengths (400–600 nm), are eliminated [58,62]. These melt pool signatures are captured by using in situ sensing sensors, which are discussed in the next section.

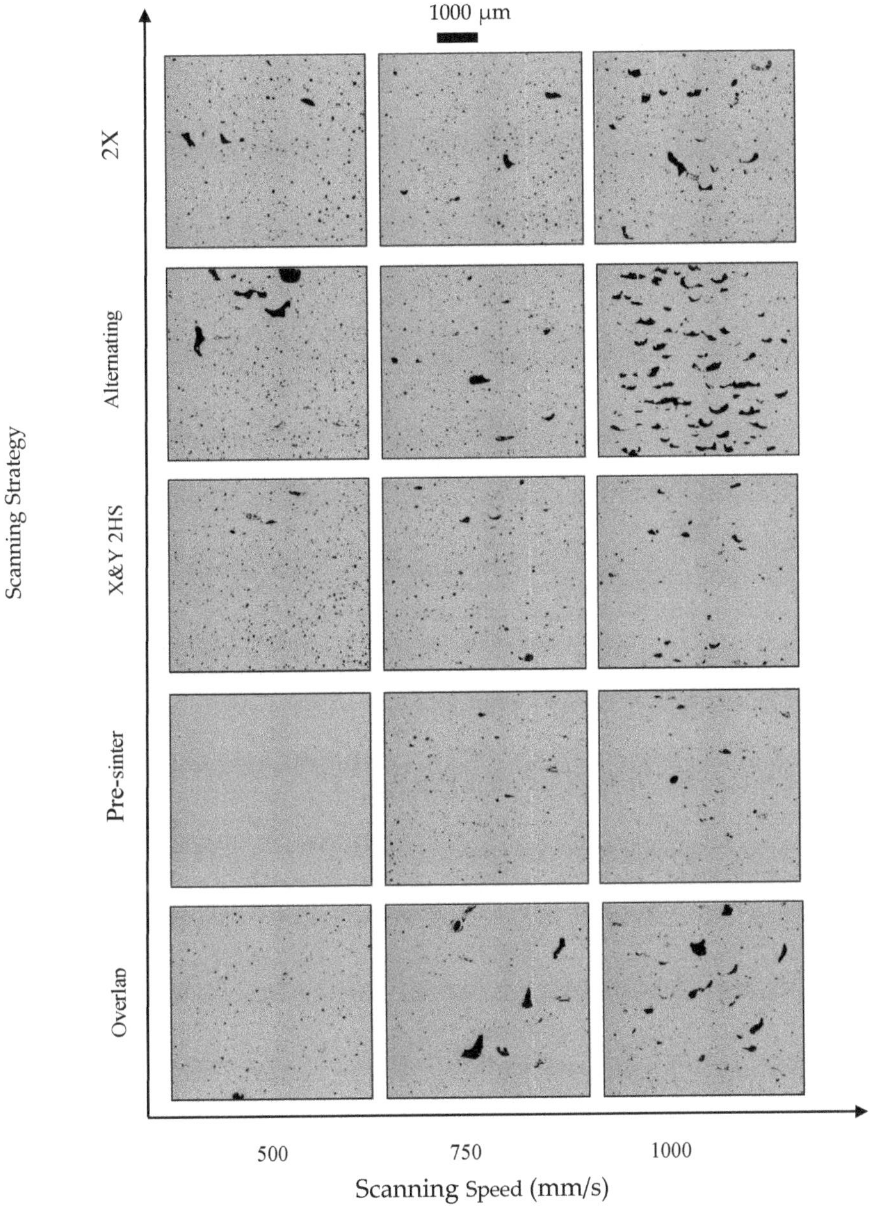

Figure 7. Effect of scanning strategies and scan speed on the porosity % of AlSi10Mg parts [30]. Reused under Creative Commons Attribution License.

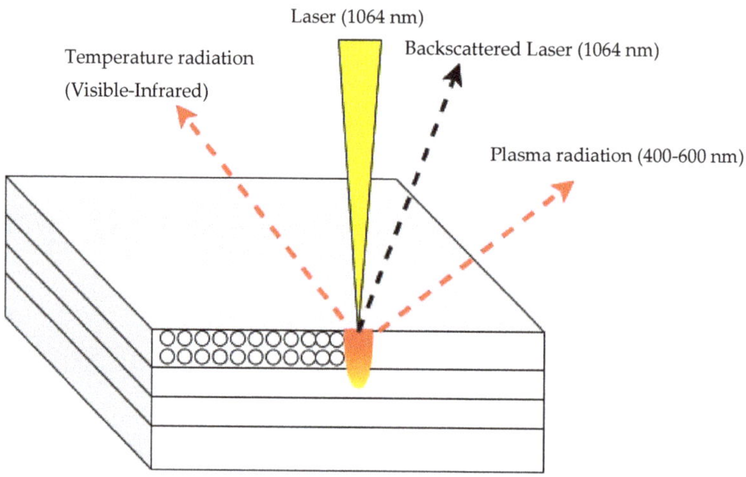

Figure 8. Illustration of types of emitting radiation from melt pool [58].

6. In Situ Sensing Devices

As discussed in the last section, capturing the melt pool signatures is essential to monitor the process. Therefore, in this section, we review the working principle of the most used devices or sensors for in situ sensing of melt pool signatures in the L-PBF process. Generally, two entities related to the melting pool are measured, i.e., melt pool morphology (shape and size) and temperature. For temperature measurements, there are two types of sensors available, i.e., photon detectors, also called non-contact sensors, and thermal detectors, referred to as contact detectors. As can be seen in Figure 9, the thermal detector response is constant with the wavelength (radiations emanating from the melt pool), whereas the photon detector response was greatly influenced by the wavelength of the light. The linking of the photon detectors output to the melt pool temperature will be discussed in Section 7. Besides monitoring the melt pool signatures, other process signatures such as powder bed spreading and part geometry are also studied using different sensor techniques, such as ultrasonic, line profiling, optical coherence tomography, etc., as reported in references [63–65].

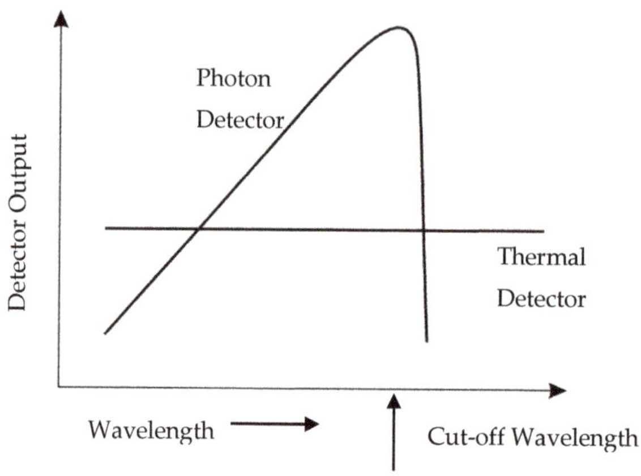

Figure 9. Wavelength response of detectors.

6.1. Non-Contact Temperature Measurement Devices

Most of the studies involve the monitoring of the temperature field or electromagnetic radiations from the melt pool both in a co-axial/on-axis and off-axis manner. The melt pool signatures capture the most invaluable information such as the temperature profile of the melt pool and the morphology of the melt pool that directly influences the quality the part [66–68]. Electromagnetic emissions are associated with the melt pool and plasma plume [27]. Therefore, to capture electromagnetic signatures, mostly non-contact measurement sensors are used. Broadly, these sensors can be classified as spatially integrated, i.e., photodiodes and pyrometers; spatially resolved, i.e., cameras; spectrally resolved, i.e., spectrometers [62,69]. Regardless of the sensors, the essential requirement for AM applications is the ability of the sensor to capture the rapid melt pool dynamics. Below, we discuss the working mechanism of the two most common photodevices used in the L-PBF process.

6.1.1. Photodiodes

Spatially integrated, single-channel detectors such as photodiodes are semiconductor devices that convert light (radiation) into electric current. The basic principle of the photodiode or photodetector is that when photons of sufficient energy (radiation) hit the active area of the device, it generates an electron-hole pair, which changes the effective resistance or conductance of the detector. A bias voltage is applied to collect the charge carriers, and the signal (photocurrent) is measured across a load resistor RL, as shown in Figure 10. The photocurrent is directly proportional to the intensity of the photons hitting the active area of the device. The intensity of photons is directly linked to the melt pool temperature. Photodiodes have the advantages and disadvantages of reducing the signal to a single data point, i.e., photovoltage corresponding to the amount of light hitting the photodiode. Reducing the signal to a single point relates to data shrinkage, which makes it easier to post-process. Low cost, high sensitivity, and high sampling rates (typically ~50 kHz) make them suitable for in situ sensing of the L-PBF process. The detection range of these devices is limited to over a range of the spectrum; for example, germanium (Ge) and InGaAs-based photodiodes are used for visible to infrared light detection. In contrast, silicon-based photodiodes are used for the UV and visible spectrum [70–73].

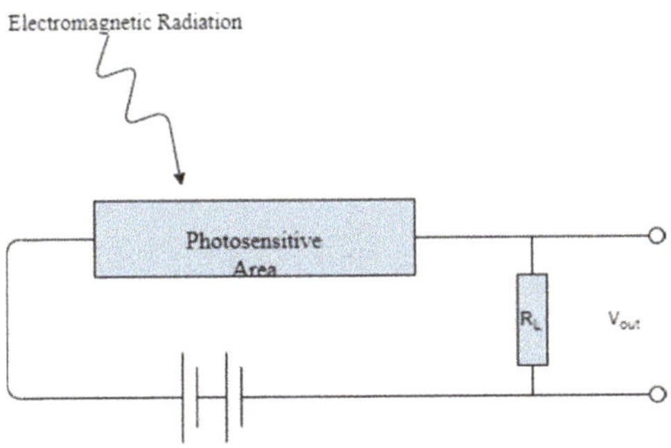

Figure 10. Schematic of working principle of PIN photodiode [70].

6.1.2. Digital Cameras

The digital cameras, also called spatially resolved sensors, are an array of photodiodes where each photodiode is labeled as a pixel. The different types of photodiodes with their detection wavelength range are given in Figure 11. Each pixel detects the photon coming from the melt pool and converts it into photocurrent. Afterward, signal processing is done to extract a picture which corresponds to the temperature fields of the melt pool. Based on the detection range of the devices, it can be classified either as a regular digital camera that has detection range in the visible spectrum or a thermal camera, which has detection range in the infrared spectrum. Mostly, for monitoring melt pool signatures, thermal cameras are preferred with a detection range over 700 nm. Based on the signal processing method, the digital cameras can be classified into two categories, i.e., charge-coupled device (CCD) and complementary metal-oxide-semiconductor (CMOS). In a CCD camera, all the pixel information is processed by the single circuit, whereas, in a CMOS camera, each pixel has its processing circuit. Therefore, the CMOS cameras have better signal processing speed, but the capture area is limited compared to a CCD camera. The main disadvantage of using cameras for in situ sensing of the L-PBF process is the speed of the process, which requires a high frame rate camera that results in an enormous size of data that are difficult to post-process [74–76].

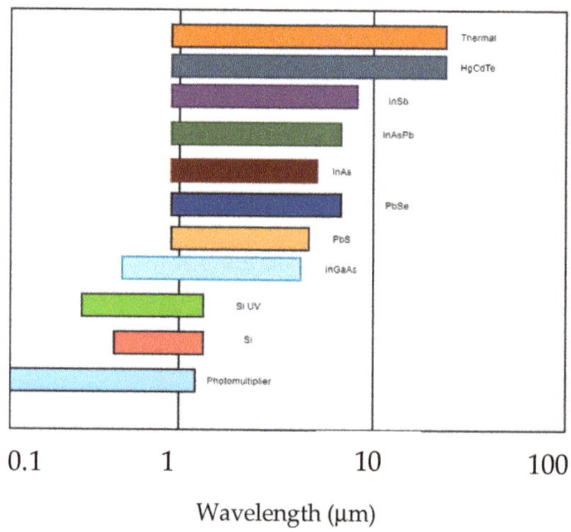

Figure 11. Wavelength sensitivity of common detectors.

6.2. Contact Temperature Measurement Devices

Another category of the devices or sensors used for monitoring the temperature fields in the L-PBF process is direct contact-based measurement devices such as thermocouples. The thermocouple is an electric device that produces a temperature-dependent voltage due to thermoelectric effect. The working principle of the thermocouple is based on the Peltier effect. The thermocouple consists of two electric conductors joined together to form two junctions. If junctions are attached to the surface at different temperatures, an electromotive force (EMF) is generated due to the temperature gradient between the two junctions of the circuit; then, total EMF can be calculated based on the temperature of the junctions and the properties of the electric conductors used. A point of measurement is connected to one of the junctions in the circuit, also called a hot junction. Another body of known temperature is connected to the other junction, also called a cold junction or a reference junction. A voltmeter measures voltage or current generated due to the temperature gradient between the two junctions. This voltage or current is then converted into the temperature [79,80]. Thermocouples can be used to monitor the overall heat flow of the part while printing. However, one limitation of the thermocouple is that it requires physical contact to measure the temperature, which is difficult in the L-PBF process and often requires extra preparation such as drilling holes or slots. Some of the researchers have monitored the L-PBF process using thermocouples, which can be found in the reference article [71].

7. Temperature Correlation

One of the key advantages of using photodiodes and digital cameras is the possibility to monitor the melt pool temperature fields or process signatures without any physical contact with the target surface. It provides the freedom to monitor the processes like AM, in which it is challenging to access the target surface. Another advantage of these sensors is the possibility to detect high temperatures such as 2000 °C. Although these sensors provide great flexibility and easy detectability, calibration of these sensors properly is a vital issue. The main principle of the sensors is to generate a photocurrent or photovoltage based on the emitted radiation from the melt pool. Therefore, it requires much understanding about the process and complex thermophysical phenomena occurring in that short interval of time. Usually, the melt pool is compared with the black body phenomenon to compute the actual temperature. However, in reality, the melt pool does not behave as an ideal black body rather than a real body. In addition, the emissivity of the material concerning different temperatures is not known. Therefore, a better understanding of black body phenomena is inevitable. Here, we present an insight into black body radiation and their correlation with the melt pool physics in L-PBF.

All bodies above absolute zero (0 K) emit thermal radiation due to the thermal vibrations of atomic particles. The quantity of the power emitted is expressed by the radiative intensity. The total energy is distributed across the electromagnetic spectrum. The emission intensity and shape of the distribution is dependent on the temperature. Conversely, the temperature of the body can be calculated by the quantity and spectral distribution of its emitted energy across the spectrum.

Firstly, a black body is a physical body that absorbs all the incident electromagnetic radiation and emits electromagnetic radiation called black body radiation. The radiation emitted by the black body is dependent on the temperature alone, not on body shape and composition. For an ideal black body at a specific temperature (T), spectral radiance (*B*) is given by Planck's law (Equation (2)), where h is Planck's constant; *c* is the speed of light; K_B is the Boltzmann constant.

$$B = \frac{2hc^2}{\lambda^5}\left(\exp\left(\frac{hc}{k_B \lambda T}\right) - 1\right)^{-1} \tag{2}$$

Figure 12 shows the spectral radiance of ideal black body radiation at different temperatures. It can be seen that the radiant intensity increases with increasing temperature at all wavelengths and for all temperatures. For any given temperature, the radiant intensity reaches a maximum at a specific wavelength, and there are two wavelengths where the radiant intensity is equal for intensities less than the maximum or peak intensity. The wavelength at maximum radiant intensity decreases with an increase in temperature (Wien's displacement law) [70].

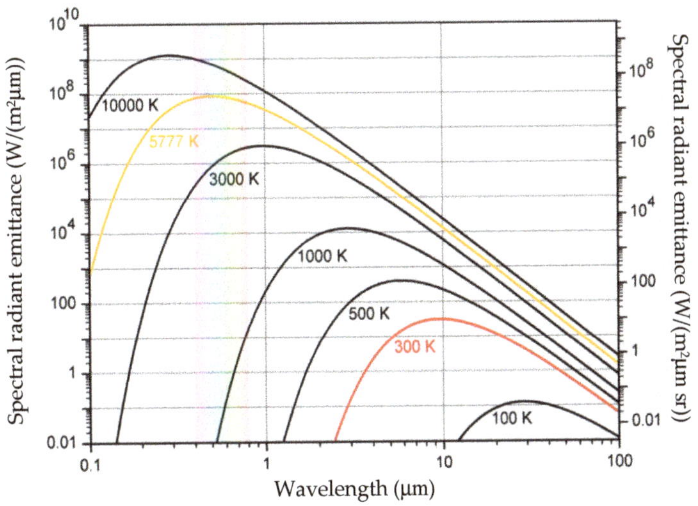

Figure 12. Spectral radiance of an ideal black body at different temperatures [70].

Nevertheless, real surfaces do not emit radiations like an idealized black body. The fraction of the emitted radiation from the real body is relative to a black body, which is known as spectral emissivity. However, spectral emissivity is temperature and wavelength-dependent. Therefore, it is challenging to obtain spectral emissivity of every material for each temperature and wavelength. Therefore, the direct application of Planck's law for temperature correlation cannot be used. Instead, to overcome this challenge, there are three possibilities which are as follows:

- For a single measurement, assume the emissivity and calculate the temperature (spectral method).
- Make two measurements at different wavelengths and assume a relationship between the emissivities at each wavelength and calculate a single temperature (ratio method).
- Make multiple measurements at different wavelengths and assume some functional form of the emissivity and find the best fit to the temperature and emissivity (multispectral method).

Mostly, the ratio method is used to calculate the actual temperature. In the ratio method, the two intensity measurements (I_1 and I_2) at two different wavelengths λ_1 and λ_2 are calculated. The ratio of intensities I_1 and I_2 are taken to calculate the temperature as given by Equation (3).

$$\frac{I_1}{I_2} = \frac{A_1 \epsilon_1 \lambda_2^5}{A_2 \epsilon_2 \lambda_1^5} \exp\left(\frac{hc}{k_B T}\left(\frac{1}{\lambda_2} - \frac{1}{\lambda_1}\right)\right) \tag{3}$$

The difference between the two wavelengths (λ_1 and λ_2) chosen is very small so that it is assumed that the emissivities (ϵ_1 and ϵ_2) are equal. Once the constants A_1 and A_2 are known, the calibration curve can be calculated, which relates the intensity ratio to the temperature of the object irrespective of the emissivity. Therefore, the choice of the two wavelengths is very critical. Because too close wavelengths can lead to low signal and noise ratio from the sensors, i.e., the measurement will be dominated by the noise. If too distinct wavelengths are chosen, the assumption of equal emissivities will no longer be valid. Most of the engineering materials have a melting point above 1000 K and vaporization temperature around 3000 K. Therefore, most of the monitoring devices have detection ranges in the visible-near infrared region. Typically, for calibration of the sensors and to find the values of the constants A_1 and A_2, the tungsten halogen light source with known spectral radiance is used. The halogen lamp is placed inside the build chamber, and the spectral radiance measured by the sensors is then compared with the radiance obtained by the calibrated spectrometer. Then, the A_1 and A_2 values are computed [70].

Moylan et al. [77] also used a multispectral method to calculate the imaged temperature from captured infrared images using a hyperspectral camera in the L-PBF process. The imaged temperature can be calculated by using this method. It shall be noted that the imaged temperature is different from the true temperature, as the true temperature is highly dependent on the emissivity of the material, which is subjected to the temperature.

8. Overview of ML Techniques

In the L-PBF process, the quality of the final part and the reliability of the whole process are highly dependent on various process parameters, as discussed in Sections 3 and 4. The above-discussed process parameters incorporate inconsistency in the part quality, which is difficult to monitor. The inconsistency during the process can be effectively controlled and monitored by using in situ monitoring systems. However, there are numerous problems associated with monitoring systems, which hinders the full exploitation of these systems. Firstly, the size of the data, real-time processing, and storage of the data are significant challenges. For example, consider a laser printing a part with geometrical tolerances of ±100 μm with a scanning speed of 100 mm/s. The sensors will have a data collection rate of ~10 kHz, i.e., one data point for every 10 μm to achieve a spatial resolution of 10 μm. Most of the researchers have reported a sampling rate of 50 kHz and used more than two sensing devices, such as a digital camera, photodiode, or combination of both. Therefore, if a single data value is stored in binary fashion, it occupies 32 bits. With a 50 kHz data sampling rate and with five sensing devices, the data will be stored at 1 MB/s. Therefore, if the machine runs for three days, it will generate at least 233 GB of data. The post-processability of this dataset and storage is difficult [78]. Secondly, linking the obtained in situ sensing data to the quality of the product is a significant challenge. In recent years, the research community is working on integrating ML methods into AM.

Machine learning (ML) is a subset of Artificial Intelligence (AI) that can learn and improve in an automated fashion from experience without being explicitly programmed. The process of learning starts with data observation and data pattern recognition to allow algorithms to learn automatically and adjust actions accordingly. The main aim of the ML is to learn and improve from data without human intervention. The ML algorithms are used in AM for design and process optimization, in situ monitoring, inspection and validation, and cost estimation. The data used for training the ML algorithms decide the effectiveness of the ML approach. In other words, the ML algorithms are as good as the training dataset [79].

ML algorithm can be broadly classified into three categories: supervised learning, unsupervised learning, and reinforcement learning. In supervised machine learning, the algorithm is trained with a labeled dataset. The labeled dataset is the dataset in which the output is known for the input variables. Therefore, the model is trained based on inferring the functional relationship between input and output variable. Supervised learning can be used for regression and classification problems. For instance, classifying the part quality as "good" and "bad" is a classification problem, whereas predicting the porosity level or mechanical properties of the part is a regression problem. In unsupervised learning, the output labels are not known for the input training dataset. The algorithm separates the training dataset into different clusters based on the relationship among input data. Unsupervised learning is useful in clustering problems such as detecting anomalous conditions. The third ML category, called reinforcement learning (RL), is learning by interacting with an environment. The RL algorithm learns from the consequences of its actions, and it selects its actions based on its experience and new choices, which is trial and error learning. The RL algorithm receives a numerical reward which encodes the success of an action's outcome. The RL algorithm seeks to learn to select actions that maximize the accumulated reward over time. Applications of the RL algorithm are game theory and self-driving cars. The list of various ML algorithms is shown in Figure 13 [80].

There is another type ML algorithm called "semi-supervised learning", which combines unsupervised and supervised learning. Semi-supervised learning combines the small or limited amount of the labeled dataset with a large amount of the unlabeled dataset during training the algorithm [81]. This type of learning is beneficial in AM applications, where obtaining the labeled dataset is a very challenging task. The decision for using a type of ML approach is based on the available dataset, and specific output preferred. The use of the abovementioned ML approaches for monitoring the L-PBF process is discussed in the next sections.

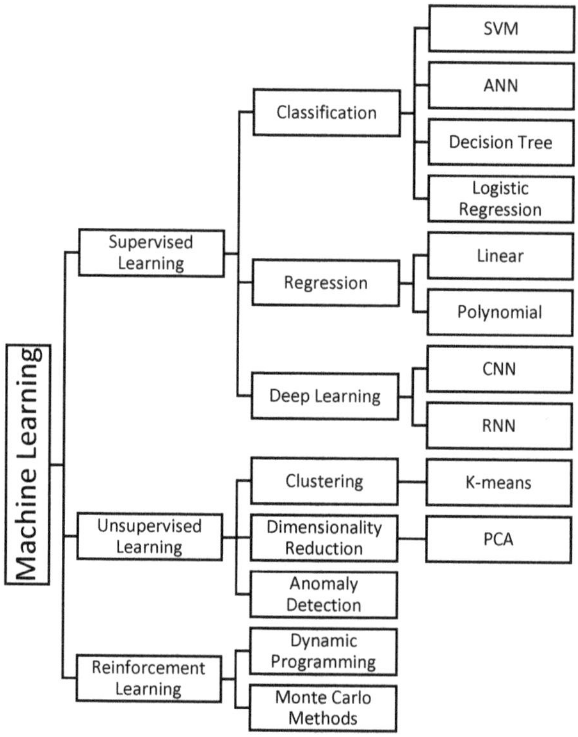

Figure 13. Schematic of machine learning algorithms.

9. Problem with ML Techniques

Although there are tremendous possibilities of the ML approach in the AM industry, there are a few limitations or problems associated with the AM methods that can have a deterministic effect on the ML outcomes.

1. Size of data: ML approaches are data-driven methods which require a sufficient amount of data for making accurate predictions. The amount of training data needed increases exponentially with a number of input variables. However, in some applications such as AM, data acquisition and labeling the training data is extremely difficult and expensive. For example, it is challenging to obtain a large amount of data for particular defects such as keyhole and lack of fusion in the L-PBF process. Thus, using neural networks that require a copious amount of data for defect detection can be a challenging task.
2. Lack of knowledge for feature extraction: Selecting good features that can capture the distinction in the dataset is most vital. Sometimes, it is very tricky to identify and extract these features. For example, it is challenging to identify the features that will distinguish an unstable melt pool signal from a stable melt pool signal in photodiode data.
3. Lack of labeled data: As discussed earlier, the supervised learning ML approach requires a labeled dataset for training. Therefore, a better understanding of the acquired data is needed to label it. However, sometimes it is challenging to label the data accurately and effectively. For example, in the case of melt pool in situ monitoring data, it is very tricky to identify the "acceptable" melt pool profile from the "unacceptable" melt pool profile. There are high chances of human error and misclassification. It can affect the accuracy of the whole model.

4. Machine learning training problems: Overfitting and underfitting are the most common problems that can occur while training machine learning models. Overfitting occurs when the model tries to fit every data point precisely in the training dataset. It makes the model vulnerable to noise and outliers. Usually, overfitting is most likely with non-parametric and non-linear models that are more flexible while learning a target function. For example, the Decision tree (DT) is a nonparametric algorithm that is very flexible and subjected to overfitting training data. Overfitting problem can be eliminated by running a tree after it has learned, to remove some of the detail it has picked up. Underfitting occurs when the model is unable to extract a reasonable relationship among training data points. An underfit model will perform poorly on training data and will not be able to generalize to new data. Drop out and regularization techniques can be used to overcome underfitting and overfitting problems [82,83].

In the next sections, we discuss the class-wise (supervised, unsupervised, semi-supervised) ML approaches used for in situ monitoring of the L-PBF process.

10. In Situ Data Processing Using ML

In this section, post-processing of the in situ monitoring data is presented as per the used ML approach classes. This section mostly focuses on in situ defects related to melt pool signatures and the powder layer spread. The use of the ML approach in other domains such as geometric control, cost estimation, process optimization, etc., is not covered in this review.

10.1. Supervised Learning Approach

Supervised learning is one of the most commonly used ML approaches for detecting the defect in the L-PBF process. The first step in supervised learning is to prepare a labeled training dataset. Labeling the in situ data is the most challenging and error-prone task, which can affect the overall accuracy of the supervised ML model. Therefore, labeling the in situ data, such as IR images, photodiode signals, pyrometer signals, can be done via making a statistical comparison between in situ data and CT images, which captures the ground truth labels. The same can be used for labeling the different clusters obtained by unsupervised learning. Gobert et al. and Petrich et al. [84,85] used a 36.3-megapixel digital single-lens reflex (DSLR) CCD camera (Nikon D800E) mounted inside the build chamber of EOS M280 (L-PBF system) to detect discontinuity defects such as overheating, pores and unmolten powder. The ground truth bales were extracted by comparing the post-CT scan data with captured in situ images. The labeled dataset was then used to train supervised ML approaches such as support vector machine (SVM) and neural network (NN). Imani et al. [86] used ML techniques like SVM, K-NN, and NN to find the process parameters such as hatch distance, laser power, and scanning speed that likely to produce more porous parts. The link between process parameters and the location, size, and frequency of the pores was studied. Aminzadeh et al. [87] used an 8.8 megapixel USB Digital Camera with high focus lenses to capture the post-printed images of every layer in L-PBF. They used Bayesian inference to detect the porosity and quality of the final part. The training dataset was labeled as "defective" and "non-defective" based on visual inspection. The frequency-domain features were extracted, and a trained Bayesian classifier was used to predict the quality of the part. Zhang et al. [88] used a high-speed camera to capture process images such as melt pool, plume, and spatters. These images were used to train SVM classifiers and CNN to detect the process anomalies. The CNN model showed a higher accuracy of 92.7% compared to the SVM classifier, which showed an accuracy of 90.1%.

Baumgartl et al. [89] used a convolutional neural network (CNN) to detect defects from thermographic images. CNN showed an average accuracy of 96.80%. The thermographic images were captured using an infrared camera, which was placed above the build chamber with an angle of 60° to the build plate. Delamination and spatter defects were successfully identified and cross-validated with the CT scan. However, the model does not identify other defects, such as balling, lack of fusion, and keyhole.

Scime et al. [90] used a one megapixel Photron FASTCAM Mini AX200 high-speed camera in the visible range to capture melt pool images. The machine learning technique, called Scale Invariant Feature

Transforms (SIFT), was used to extract features from the captured melt pool signatures. The bag-of-words (or Keypoints) were used to obtain a scale-agnostic description of melt pool morphology. The melt pool morphology represents the shape of the melt pool, spatter, and vapor plume. With the ML approach, the author classified the individual melt pool morphologies into four categories, such as desirable, balling, under-melting, and keyhole porosities. A supervised classification ML technique called support vector machine (SVM) was used to train and test the model. The percentage of the melt pool morphologies linked to process parameters such as laser power and scan speed was obtained.

In another study, Scime et al. [91] used a computer vision algorithm to predict the percentage of types of defects present in the part based on the powder bed images. Scime et al. extracted the regions of the images and classify such as anomaly-free, recoater hopping, recoater streaking, debris, superelevation, part failure, and incomplete spreading. These categories were used as the fingerprints for the computer vision algorithm. A percentage of the particular defect in the respective layer was classified. The approach did not use melt pool signatures and instead used powder bed spreading as their main signature. Repossini et al. [92] used a high-speed camera to capture spatter images and used them as an indicator for process stability. They have developed a supervised learning logistic regression model to classify different energy density conditions to various quality states using spatters as a descriptor.

Some researchers also studied the acoustic signals to detect the defects in the L-PBF process using a supervised learning ML approach. For example, Ye et al. [93] used deep belief networks (DBN) to analyze the acoustic signals to detect defects such as balling, keyhole, and cracking. Shevchik et al. [94] used Bragg grating acoustic sensors to capture the acoustic signals during melting, solidification, and spattering processes. The time-domain captured signal was transformed into the frequency domain using the wavelet transformation technique. The spectral CNN was used to identify the particular defect signals. The presented sCNN model showed an accuracy of 83–89%.

10.2. Unsupervised Learning Approach

The use of unsupervised machine learning is limited in the field of the L-PBF process. However, unsupervised can be useful when obtaining the labeled training dataset is not possible. Another advantage of using unsupervised learning compared to supervised learning is that it does not require human interaction to the label training dataset. Grasso et al. [10] demonstrated the use of the T-mode Principal Component Analysis (PCA) for image data obtained by an Olympus^{TM} I-speed 3 camera mounted outside the build chamber at an angle of 40° w.r.t build plate to define a spatial statistical descriptor and to detect local over-heating phenomena along the scan path by analyzing the intensity profile of each pixel. These local hotspots correspond to the regions where slow cooling occurred due to variation in heat flux; it can be observed at sharp corners and overhang structures. K-means clustering learning was applied to the spatially distributed PCA descriptors to detect a defect in an automated fashion. Although an iterative updating of the k-means clustering based on the new dataset allowed to detect defects in complex shapes, the approach is dependent on the comparisons between data collected at different layers or on the signal from successful builds of the same geometry, which limits its applicability to other situations. Grasso et al. [95] used an off-axis IR camera to monitor the plume of zinc material to design a data-driven and automated approach for process monitoring. The region of interest (ROI), which is the plume, was extracted from the IR images. A multivariate control-charting method was proposed to monitor the ROI features, i.e., mean intensity and area of the plume, to detect the deviation from the stable plume patterns. However, in this work, the author did not take into consideration the morphology of the melt pool.

10.3. Semi-Supervised Learning

Okaro et al. [96] proposed another perspective of the ML approach for data treatment and predicted the quality of the parts based on their mechanical property as a descriptor. Key features were extracted from the photodiode signals, and a semi-supervised classification algorithm called "Gaussian

Mixture Model-Expectation Maximization (GMM-EM)" was applied to classify the samples as "faulty" and "acceptable" based on the ultimate tensile strength of the tensile bars. This approach showed the possibility for automatic certification of the L-PBF parts based on their mechanical properties. However, the method of Okaro et al. does not illustrate any possibility to detect the defects in the parts, which is the cause of inferior mechanical properties.

Yuan et al. [97] used a high-speed 1 kHz video camera, with its focal point moving along with the laser spot, to acquire in situ video data to monitor scan track width and continuity. Image processing algorithms were used to label a small amount of data using structured light measurements. The labeled in situ videos were used to train a semi-supervised CNN model based on the temporal ensemble method, with a small amount of labeled data and a large amount of unlabeled data. Yuan et al. showed that the semi-supervised CNN approach performs better than the supervised CNN approach. They also studied the average measuring time for 100 videos, which was 1.4 ms and concluded that the trained CNN approach is adequately fast to monitor in real-time. The summary of the discussed literature is presented in Table 1.

Table 1. List of the literature on the processing of L-PBF in situ data using machine learning approaches.

Reference	Data Type	Monitoring Quantity	ML Technique
	Supervised Learning Approach		
Gobert et al. [84] and Petrich et al. [85]	Images of post-printing	Defects such as overheating, pores and unmolten powder	SVM, NN
Imani et al. [86]	process parameters such as hatch distance, laser power, and scanning speed	Porosity level	K-NN, SVM, NN
Aminzadeh et al. [87]	Images of post-printing	Porosity	Bayesian inference
Zhang et al. [88]	Melt pool, spatters and plume images	Process anomalies	SVM and CNN
Baumgartl et al. [89]	Thermographic images	Delamination and spatter defects	CNN
Scime et al. [90]	Melt pool images	Types of melt pool shapes	SIFT, SVM
Scime et al. [91]	Powder bed images	Powder bed defects	Computer vision
Ye et al. [93]	Acoustic signals	Defects such as balling, keyhole, and cracking	DBN
Shevchik et al. [94]	Acoustic signals	Defect signals	sCNN
	Unsupervised Learning Approach		
Grasso et al. [10]	Images	Overheating defect	T-Mode PCA, K-means
	Semi-supervised Learning Approach		
Okaro et al. [96]	Photodiode signal	Defective parts based on tensile strength	GMM-EM
Yuan et al. [97]	Video	Scan track width and continuity	Semi-supervised CNN

The final aim of in situ monitoring is to improve the overall reliability of the process via a feedback control loop. This will significantly reduce the material waste and time needed to qualify the part for quality assurance. Yao et al. [98] developed a closed-loop feedback control system for real-time monitoring of the L-PBF process. Firstly, the defect level of each printed layer was calculated based on irregular and non-homogeneous patterns using multifractal analysis. Secondly, the composite index was decided for each porosity level based on Hotelling T^2 statistics from the multifractal spectrum. Finally, the stochastic dynamics of the layer to layer defect conditions were modeled as a Markov process to determine an optimal control policy.

11. Commercial Monitoring Systems

Due to the increasing demand for L-PBF systems in manufacturing industries, the commercial suppliers of L-PBF systems are now equipped with the inbuilt in situ sensing modules. The monitoring system layout, detection range, and types of sensors differentiate different commercial supplier monitoring systems. There is no common consensus on the detection range of the melt pool emissions among different suppliers. The reason to choose a specific detection range is protected under the confidentiality clause by the supplier. Therefore, each module has its disadvantages and advantages. For example, the EOSTATE Optical Tomography (OT) module detects the light emitted from the met pool for every layer. The EOSTATE Exposure analysis tool provided by EOS GmbH uses a statistical algorithm to detect so-called "hotspots" and "cold spots", which are the regions of high and low intensity compared to rest of the layer.

Similarly, Melt Pool Monitoring (MPM) provided by SLM solutions captures the melt pool emissions in the near-infrared region by using two photodiodes with different spectral ranges. Then, captured thermal emissions can be studied in the software module supplied by SLM solutions in a layer by layer fashion. The Layer Control System (LCS), also supplied by SLM solutions, uses the visible range camera to monitor the quality of the powder bed spread. After the powder bed spreading, an automatic algorithm processes the captured image in real-time and automatically activates the recoating operation in case of inhomogeneous powder spread. Renishaw supplied the InfiAM module composed of MeltView, which captures the in situ data in a wide spectral range and LaserView to monitor the part build and laser power, respectively. It provides the captured data in 2D and 3D view as the build progresses. Table 2 summarizes the in situ sensing equipment with their sensing process signature capabilities. It shall be noted that the author's aim is not to comment on the capabilities of the available commercial systems as the authors have not personally used all these systems.

Table 2. List of commercial in situ sensing modules.

Module Name.	Developer	Monitored Quantity	In Situ Sensing
QM meltpool 3D [99]	Concept Laser	Melt pool (area and intensity)	Co-axial photodiodes (co-axial camera also available in research version)
EOSTATE MeltPool	EOS	Melt pool	Co-axial and off-axial sensors
EOSTATE [100] Exposure OT [100]	EOS	Thermal map over the entire powder bed	Off-axis camera
Melt Pool Monitoring (MPM) system [101]	SLM Solutions	Melt pool	Co-axial pyrometer
Layer Control System (LCS) [101]	SLM Solutions	Powder bed	Off-axial camera
InfiAM [102]	Renishaw	Melt pool	Co-axial photodiodes
Truprint Monitoring [103]	Trumpf	Melt pool	Co-axial photodiodes (beta version)
Truprint Monitoring [103]	Trumpf	Powder bed and part geometry	Off-axial camera
	SISMA	Powder bed	Off-axial camera
PrintRite3D [104]	B6 Sigma, Inc.	Different monitoring equipment	Set of co-axial and off-axial sensors available

12. Conclusions

From the literature available, it is evident that the in situ sensing of the L-PBF process is still in the infancy stage. Much of the research conducted so far focuses on developing in situ sensing systems to understand the process and less concentrated on detecting defects in real-time. The ultimate goal of in situ sensing in the L-PBF process is to develop a closed feedback control system to improve and control the quality assurance in real-time. Understanding the process signatures and their link to the potential defect in the final part is the first step towards this. Commercial systems equipped with monitoring systems work on a black box phenomenon, and the critical information obtained during the process is protected by the manufacturer. The large size of the data generated for a particular job is another limitation of these systems. Thus, there is a huge demand for other advanced machine learning models for treating and extracting crucial information from the available captured data. To date, only a few researchers have used a machine learning approach to counter this problem.

The complexity of the L-PBF process is a significant challenge to understand all the phenomena occurring due to controllable and uncontrollable parameters. Therefore, monitoring each stage of the process, such as powder health, powder spreadability, build chamber inert gas flow, oxygen content, melt pool stability, etc., are closely interlinked. Therefore, linking the different process states such as powder bed spreading, layer heatmap, and melt pool stability can enhance the quality control in real-time. Real-time monitoring will also limit the use of other quality checks, such as non-destructive testing, which costs a lot of time and money. The improved reliability and repeatability of the L-PBF process will allow risk-averse industries to adopt the process entirely.

Author Contributions: Conceptualization and methodology, P.Y., E.L., O.R., C.A., E.L.G.; investigation, E.L., O.R., C.A.; writing—original draft preparation, P.Y.; writing—review and editing, P.Y., E.L., O.R., C.R.; visualization, P.Y.; supervision, E.L., O.R., C.R., E.L.G.; All authors have read and agreed to the published version of the manuscript.

Funding: This work was conducted as part of the "ENABLE" project funded by the European Union's Marie Sklodowska-Curie Actions (MSCA) Innovative Training Networks (ITN) H2020-MSCA-ITN-2017 under the grant agreement Number 764979.

Acknowledgments: The authors would like to thank the T-ADD team at SIRRIS for providing valuable suggestions and seminars related to the L-PBF process.

Conflicts of Interest: The authors declare no conflict of interest.

Abbreviations

AI	Artificial Intelligence
AM	Additive Manufacturing
ANN	Artificial Neural Networks
ASTM	American Society for Testing and Materials
CCD	Charge-Coupled Device
CMOS	Complementary Metal–Oxide–Semiconductor
CNN	Convolutional Neural Network
DBN	Deep Belief Networks
DMLS	Direct Melt Laser Sintering
DT	Decision Tree
EMF	Electromotive Force
GMM-EM	Gaussian Mixture Model- Expectation-Maximization
KNN	k-Nearest Neighbors
LOF	Lack of Fusion
L-PBF	Laser-Powder Bed Fusion
ML	Machine Learning
NN	Neural Network
PCA	Principal Component Analysis
RL	Reinforcement Learning
SIFT	Scale Invariant Feature Transforms
SLM	Selective Laser Melting
SVM	Support Vector Machine
TGM	Temperature Gradient Mechanism

References

1. ASTM Standard. *ISO/ASTM 52900: 2015 Additive Manufacturing-General Principles-Terminology*; ASTM F2792-10e1; ASTM Standard: West Kanshohoken, PA, USA, 2012.
2. Mani, M.; Feng, S.; Lane, B.; Donmez, A.; Moylan, S.; Fesperman, R. Measurement Science Needs for Real-Time Control of Additive Manufacturing Powder Bed Fusion Processes. 2015. Available online: https://www.researchgate.net/publication/279178288_Measurement_Science_Needs_for_Real-time_Control_of_Additive_Manufacturing_Powder_Bed_Fusion_Processes (accessed on 17 June 2020).
3. Galy, C.; Le Guen, E.; Lacoste, E.; Arvieu, C. Main defects observed in aluminum alloy parts produced by SLM: From causes to consequences. *Addit. Manuf.* **2018**, *22*, 165–175. [CrossRef]
4. Tapia, G.; Elwany, A. A review on process monitoring and control in metal-based additive manufacturing. *J. Manuf. Sci. Eng.* **2014**, *136*, 060801. [CrossRef]
5. Khoda, A.; Ozbolat, I.T.; Koc, B. Designing heterogeneous porous tissue scaffolds for additive manufacturing processes. *Comput. Aided Des.* **2013**, *45*, 1507–1523. [CrossRef]
6. Huang, Q.; Zhang, J.; Sabbaghi, A.; Dasgupta, T. Optimal offline compensation of shape shrinkage for three-dimensional printing processes. *Iie Trans.* **2015**, *47*, 431–441. [CrossRef]
7. Everton, S.K.; Hirsch, M.; Stravroulakis, P.; Leach, R.K.; Clare, A.T. Review of in-situ process monitoring and in-situ metrology for metal additive manufacturing. *Mater. Des.* **2016**, *95*, 431–445. [CrossRef]
8. Spears, T.G.; Gold, S.A. In-process sensing in selective laser melting (SLM) additive manufacturing. *Integr. Mater. Manuf. Innov.* **2016**, *5*, 16–40. [CrossRef]
9. Grasso, M.; Colosimo, B.M. Process defects and in situ monitoring methods in metal powder bed fusion: A review. *Meas. Sci. Technol.* **2017**, *28*, 044005. [CrossRef]
10. Grasso, M.; Laguzza, V.; Semeraro, Q.; Colosimo, B.M. In-process monitoring of selective laser melting: Spatial detection of defects via image data analysis. *J. Manuf. Sci. Eng.* **2017**, *139*, 051001. [CrossRef]
11. You, D.; Gao, X.; Katayama, S. Review of laser welding monitoring. *Sci. Technol. Weld. Join.* **2014**, *19*, 181–201. [CrossRef]
12. Clijsters, S.; Craeghs, T.; Buls, S.; Kempen, K.; Kruth, J.-P. In situ quality control of the selective laser melting process using a high-speed, real-time melt pool monitoring system. *Int. J. Adv. Manuf. Technol.* **2014**, *75*, 1089–1101. [CrossRef]
13. Craeghs, T.; Bechmann, F.; Berumen, S.; Kruth, J.-P. Feedback control of Layerwise Laser Melting using optical sensors. *Phys. Procedia* **2010**, *5*, 505–514. [CrossRef]
14. Craeghs, T.; Clijsters, S.; Kruth, J.-P.; Bechmann, F.; Ebert, M.-C. Detection of process failures in layerwise laser melting with optical process monitoring. *Phys. Procedia* **2012**, *39*, 753–759. [CrossRef]
15. Craeghs, T.; Clijsters, S.; Yasa, E.; Bechmann, F.; Berumen, S.; Kruth, J.-P. Determination of geometrical factors in Layerwise Laser Melting using optical process monitoring. *Opt. Lasers Eng.* **2011**, *49*, 1440–1446. [CrossRef]
16. Demir, A.G.; Mazzoleni, L.; Caprio, L.; Pacher, M.; Previtali, B. Complementary use of pulsed and continuous wave emission modes to stabilize melt pool geometry in laser powder bed fusion. *Opt. Laser Technol.* **2019**, *113*, 15–26. [CrossRef]
17. Dobrzański, L.A.; Dobrzańska-Danikiewicz, A.D.; Achtelik-Franczak, A.; Dobrzański, L.B.; Szindler, M.; Gaweł, T.G. Porous selective laser melted Ti and Ti6Al4V materials for medical applications. In *Powder Metallurgy–Fundamentals and Case Studies*; InTech: Rijeka, Croatia, 2017; pp. 161–181.
18. Fischer, P.; Romano, V.; Weber, H.-P.; Karapatis, N.; Boillat, E.; Glardon, R. Sintering of commercially pure titanium powder with a Nd: YAG laser source. *Acta Mater.* **2003**, *51*, 1651–1662. [CrossRef]
19. Hebert, R.J. metallurgical aspects of powder bed metal additive manufacturing. *J. Mater. Sci.* **2016**, *51*, 1165–1175. [CrossRef]
20. Kruth, J.; Bartscher, M.; Carmignato, S.; Schmitt, R.; De Chiffre, I.; Weckenmann, A. Computed tomography for dimensional metrology. *CIRP Ann.* **2011**, *60*, 821–842.
21. Song, B.; Dong, S.; Deng, S.; Liao, H.; Coddet, C. Microstructure and tensile properties of iron parts fabricated by selective laser melting. *Opt. Laser Technol.* **2014**, *56*, 451–460. [CrossRef]
22. Karimi, P.; Raza, T.; Andersson, J.; Svensson, L.-E. Influence of laser exposure time and point distance on 75-μm-thick layer of selective laser melted Alloy 718. *Int. J. Adv. Manuf. Technol.* **2018**, *94*, 2199–2207. [CrossRef]

23. Gu, H.; Gong, H.; Pal, D.; Rafi, K.; Starr, T.; Stucker, B. Influences of energy density on porosity and microstructure of selective laser melted 17-4PH stainless steel. In Proceedings of the 2013 Solid Freeform Fabrication Symposium, Austin, TX, USA, 12–14 August 2013.
24. Slotwinski, J.A.; Garboczi, E.J. Porosity of additive manufacturing parts for process monitoring. In *AIP Conference Proceedings*; American Institute of Physics: College Park, MD, USA, 2014; pp. 1197–1204.
25. Weingarten, C.; Buchbinder, D.; Pirch, N.; Meiners, W.; Wissenbach, K.; Poprawe, R. Formation and reduction of hydrogen porosity during selective laser melting of AlSi10Mg. *J. Mater. Process. Technol.* **2015**, *221*, 112–120.
26. Sharratt, B.M. *Non-Destructive Techniques and Technologies for Qualification of Additive Manufactured Parts and Processes*; Sharratt Research and Consulting Inc.: Victoria, BC, USA, 2015.
27. Thijs, L.; Verhaeghe, F.; Craeghs, T.; Van Humbeeck, J.; Kruth, J.-P. A study of the microstructural evolution during selective laser melting of Ti–6Al–4V. *Acta Mater.* **2010**, *58*, 3303–3312. [CrossRef]
28. Das, S. Physical aspects of process control in selective laser sintering of metals. *Adv. Eng. Mater.* **2003**, *5*, 701–711. [CrossRef]
29. Gong, H.; Rafi, K.; Gu, H.; Starr, T.; Stucker, B. Analysis of defect generation in Ti–6Al–4V parts made using powder bed fusion additive manufacturing processes. *Addit. Manuf.* **2014**, *1*, 87–98. [CrossRef]
30. Aboulkhair, N.T.; Everitt, N.M.; Ashcroft, I.; Tuck, C. Reducing porosity in AlSi10Mg parts processed by selective laser melting. *Addit. Manuf.* **2014**, *1*, 77–86. [CrossRef]
31. Song, B.; Dong, S.; Liu, Q.; Liao, H.; Coddet, C. Vacuum heat treatment of iron parts produced by selective laser melting: Microstructure, residual stress and tensile behavior. *Mater. Des.* **2014**, *54*, 727–733. [CrossRef]
32. Rafi, H.; Karthik, N.; Gong, H.; Starr, T.L.; Stucker, B.E. Microstructures and mechanical properties of Ti6Al4V parts fabricated by selective laser melting and electron beam melting. *J. Mater. Eng. Perform.* **2013**, *22*, 3872–3883. [CrossRef]
33. Cunningham, R.; Zhao, C.; Parab, N.; Kantzos, C.; Pauza, J.; Fezzaa, K.; Sun, T.; Rollett, A.D. Keyhole threshold and morphology in laser melting revealed by ultrahigh-speed x-ray imaging. *Science* **2019**, *363*, 849–852. [CrossRef]
34. Liu, Q.C.; Elambasseril, J.; Sun, S.J.; Leary, M.; Brandt, M.; Sharp, P.K. The effect of manufacturing defects on the fatigue behaviour of Ti-6Al-4V specimens fabricated using selective laser melting. In *Advanced Materials Research*; Trans Tech Publications Ltd.: Kapellweg, Switzerland, 2014; pp. 1519–1524.
35. Vilaro, T.; Colin, C.; Bartout, J.-D. As-fabricated and heat-treated microstructures of the Ti-6Al-4V alloy processed by selective laser melting. *Metall. Mater. Trans. A* **2011**, *42*, 3190–3199. [CrossRef]
36. Read, N.; Wang, W.; Essa, K.; Attallah, M.M. Selective laser melting of AlSi10Mg alloy: Process optimisation and mechanical properties development. *Mater. Des.* **2015**, *65*, 417–424. [CrossRef]
37. Antony, K.; Arivazhagan, N. Studies on energy penetration and marangoni effect during laser melting process. *J. Eng. Sci. Technol.* **2015**, *10*, 509–525.
38. Gu, D.; Shen, Y. Balling phenomena in direct laser sintering of stainless steel powder: Metallurgical mechanisms and control methods. *Mater. Des.* **2009**, *30*, 2903–2910. [CrossRef]
39. Li, R.; Liu, J.; Shi, Y.; Wang, L.; Jiang, W. Balling behavior of stainless steel and nickel powder during selective laser melting process. *Int. J. Adv. Manuf. Technol.* **2012**, *59*, 1025–1035. [CrossRef]
40. Tolochko, N.K.; Mozzharov, S.E.; Yadroitsev, I.A.; Laoui, T.; Froyen, L.; Titov, V.I.; Ignatiev, M.B. Balling processes during selective laser treatment of powders. *Rapid Prototyp. J.* **2004**, 10. [CrossRef]
41. Mercelis, P.; Kruth, J.P. Residual stresses in selective laser sintering and selective laser melting. *Rapid Prototyp. J.* **2006**, *12*. [CrossRef]
42. Harrison, N.J.; Todd, I.; Mumtaz, K. Reduction of micro-cracking in nickel superalloys processed by Selective Laser Melting: A fundamental alloy design approach. *Acta Mater.* **2015**, *94*, 59–68. [CrossRef]
43. Mercelis, P. Control of Selective Laser Sintering and Selective Laser Melting Processes. Ph.D. Thesis, DentWise Division, Leuven, Belgium, 2007.
44. Carter, L.N.; Martin, C.; Withers, P.J.; Attallah, M.M. The influence of the laser scan strategy on grain structure and cracking behaviour in SLM powder-bed fabricated nickel superalloy. *J. Alloy. Compd.* **2014**, *615*, 338–347. [CrossRef]
45. Carter, L.N.; Attallah, M.M.; Reed, R.C. Laser powder bed fabrication of nickel-base superalloys: Influence of parameters; characterisation, quantification and mitigation of cracking. *Superalloys* **2012**, *2012*, 577–586.

46. Carter, L.N.; Wang, X.; Read, N.; Khan, R.; Aristizabal, M.; Essa, K.; Attallah, M.M. Process optimisation of selective laser melting using energy density model for nickel based superalloys. *Mater. Sci. Technol.* **2016**, *32*, 657–661. [CrossRef]
47. Zhao, X.; Lin, X.; Chen, J.; Xue, L.; Huang, W. The effect of hot isostatic pressing on crack healing, microstructure, mechanical properties of Rene88DT superalloy prepared by laser solid forming. *Mater. Sci. Technol. A* **2009**, *504*, 129–134. [CrossRef]
48. Jing, C.; Xin, L.; Tao, W.; Haiou, Y.; Weidong, H. The hot cracking mechanism of 316L stainless steel cladding in rapid laser forming process. *Rare Met. Mater. Eng.* **2003**, *32*, 183–186.
49. Kruth, J.-P.; Deckers, J.; Yasa, E.; Wauthlé, R. Assessing and comparing influencing factors of residual stresses in selective laser melting using a novel analysis method. Proceedings of the institution of mechanical engineers. *Part B J. Eng. Manuf.* **2012**, *226*, 980–991. [CrossRef]
50. Mugwagwa, L.; Dimitrov, D.; Matope, S.; Yadroitsev, I. Influence of process parameters on residual stress related distortions in selective laser melting. *Procedia Manuf.* **2018**, *21*, 92–99. [CrossRef]
51. Mishurova, T.; Cabeza, S.; Thiede, T.; Nadammal, N.; Kromm, A.; Klaus, M.; Genzel, C.; Haberland, C.; Bruno, G. The influence of the support structure on residual stress and distortion in SLM Inconel 718 parts. *Metall. Mater. Trans. A* **2018**, *49*, 3038–3046. [CrossRef]
52. Song, B.; Zhao, X.; Li, S.; Han, C.; Wei, Q.; Wen, S.; Liu, J.; Shi, Y. Differences in microstructure and properties between selective laser melting and traditional manufacturing for fabrication of metal parts: A review. *Front. Mech. Eng.* **2015**, *10*, 111–125. [CrossRef]
53. Townsend, A.; Senin, N.; Blunt, L.; Leach, R.; Taylor, J. Surface texture metrology for metal additive manufacturing: A review. *Precis. Eng.* **2016**, *46*, 34–47. [CrossRef]
54. Strano, G.; Hao, L.; Everson, R.M.; Evans, K.E. Surface roughness analysis, modelling and prediction in selective laser melting. *J. Mater. Process. Technol.* **2013**, *213*, 589–597. [CrossRef]
55. Fox, J.C.; Moylan, S.P.; Lane, B.M. Effect of process parameters on the surface roughness of overhanging structures in laser powder bed fusion additive manufacturing. *Procedia Cirp* **2016**, *45*, 131–134. [CrossRef]
56. Yadroitsev, I.; Bertrand, P.; Smurov, I. Parametric analysis of the selective laser melting process. *Appl. Surf. Sci.* **2007**, *253*, 8064–8069. [CrossRef]
57. Sun, K.; Peng, W.; Yang, L.; Fang, L. Effect of SLM processing parameters on microstructures and mechanical properties of Al0. 5CoCrFeNi high entropy alloys. *Metals* **2020**, *10*, 292. [CrossRef]
58. Zenzinger, G.; Bamberg, J.; Ladewig, A.; Hess, T.; Henkel, B.; Satzger, W. Process monitoring of additive manufacturing by using optical tomography. In *AIP Conference Proceedings*; American Institute of Physics: College Park, MD, USA, 2015; pp. 164–170.
59. Kurzynowski, T.; Chlebus, E.; Kuźnicka, B.; Reiner, J. Parameters in selective laser melting for processing metallic powders. In *High Power Laser Materials Processing: Lasers, Beam Delivery, Diagnostics, and Applications*; International Society for Optics and Photonics: Bellingham, DC, USA, 2012; p. 823914.
60. Hoffman, J.; Szymański, Z. Time-dependent spectroscopy of plasma plume under laser welding conditions. *J. Phys. D Appl. Phys.* **2004**, *37*, 1792. [CrossRef]
61. Hauser, C.; Childs, T.; Dalgarno, K.; Eane, R. Atmospheric control during direct selective laser sintering of stainless steel 314S powder. 1999. Available online: https://repositories.lib.utexas.edu/bitstream/handle/2152/73605/1999-030-Hauser.pdf?sequence=2&isAllowed=y (accessed on 18 June 2020).
62. Lott, P.; Schleifenbaum, H.; Meiners, W.; Wissenbach, K.; Hinke, C.; Bültmann, J. Design of an optical system for the in situ process monitoring of selective laser melting (SLM). *Phys. Procedia* **2011**, *12*, 683–690. [CrossRef]
63. Smith, R.J.; Hirsch, M.; Patel, R.; Li, W.; Clare, A.T.; Sharples, S.D. Spatially resolved acoustic spectroscopy for selective laser melting. *J. Mater. Process. Technol.* **2016**, *236*, 93–102. [CrossRef]
64. Guan, G.; Hirsch, M.; Syam, W.P.; Leach, R.K.; Huang, Z.; Clare, A.T. Loose powder detection and surface characterization in selective laser sintering via optical coherence tomography. *Proc. R. Soc. A Math. Phys. Eng. Sci.* **2016**, *472*, 20160201. [CrossRef]
65. Guan, G.; Hirsch, M.; Lu, Z.H.; Childs, D.T.; Matcher, S.J.; Goodridge, R.; Groom, K.M.; Clare, A.T. Evaluation of selective laser sintering processes by optical coherence tomography. *Mater. Des.* **2015**, *88*, 837–846. [CrossRef]
66. Wang, H. Materials' fundamental issues of laser additive manufacturing for high-performance large metallic components. *Acta Aeronaut. Astronaut. Sin.* **2014**, *35*, 2690–2698.

67. Caiazzo, F.; Cardaropoli, F.; Alfieri, V.; Sergi, V.; Cuccaro, L. Experimental analysis of Selective Laser Melting process for Ti-6Al-4V turbine blade manufacturing. In Proceedings of the XIX International Symposium on High-Power Laser Systems and Applications 2012, Nara, Japan, 26–31 August 2012; p. 86771H.
68. Yan, C.; Hao, L.; Hussein, A.; Young, P. Ti–6Al–4V triply periodic minimal surface structures for bone implants fabricated via selective laser melting. *J. Mech. Behav. Biomed. Mater.* **2015**, *51*, 61–73. [CrossRef] [PubMed]
69. Purtonen, T.; Kalliosaari, A.; Salminen, A. Monitoring and adaptive control of laser processes. *Phys. Procedia* **2014**, *56*, 1218–1231. [CrossRef]
70. Coates, P.; Lowe, D. *The Fundamentals of Radiation Thermometers*; CRC Press: Boca Raton, FL, USA, 2016.
71. Held, G. *Introduction to Light Emitting Diode Technology and Applications*; CRC press: Boca Raton, FL, USA, 2016.
72. Bi, G.; Gasser, A.; Wissenbach, K.; Drenker, A.; Poprawe, R. Identification and qualification of temperature signal for monitoring and control in laser cladding. *Opt. Lasers Eng.* **2006**, *44*, 1348–1359. [CrossRef]
73. Bi, G.; Schürmann, B.; Gasser, A.; Wissenbach, K.; Poprawe, R. Development and qualification of a novel laser-cladding head with integrated sensors. *Int. J. Mach. Tools Manuf.* **2007**, *47*, 555–561. [CrossRef]
74. Fathi, A.; Khajepour, A.; Durali, M.; Toyserkani, E. Geometry control of the deposited layer in a nonplanar laser cladding process using a variable structure controller. *J. Manuf. Sci. Eng.* **2008**, *130*, 031003. [CrossRef]
75. Zeinali, M.; Khajepour, A. Height control in laser cladding using adaptive sliding mode technique: Theory and experiment. *J. Manuf. Sci. Eng.* **2010**, *132*, 041016. [CrossRef]
76. Salehi, D.; Brandt, M. Melt pool temperature control using LabVIEW in Nd: YAG laser blown powder cladding process. *Int. J. Adv. Manuf. Technol.* **2006**, *29*, 273–278. [CrossRef]
77. Moylan, S.; Whitenton, E.; Lane, B.; Slotwinski, J. Infrared thermography for laser-based powder bed fusion additive manufacturing processes. In *AIP Conference Proceedings*; American Institute of Physics: College Park, MD, USA, 2014; pp. 1191–1196.
78. Li, Y.; Gu, D. Thermal behavior during selective laser melting of commercially pure titanium powder: Numerical simulation and experimental study. *Addit. Manuf.* **2014**, *1*, 99–109. [CrossRef]
79. Wuest, T.; Weimer, D.; Irgens, C.; Thoben, K.-D. Machine learning in manufacturing: Advantages, challenges, and applications. *Prod. Manuf. Res.* **2016**, *4*, 23–45. [CrossRef]
80. Alpaydin, E. *Introduction to Machine Learning*; MIT Press: Cambridge, MA, USA, 2020.
81. Chapelle, O.; Scholkopf, B.; Zien, A. Semi-supervised learning (chapelle, o. et al., eds.; 2006). *IEEE Trans. Neural Netw.* **2009**, *20*, 542. [CrossRef]
82. Ng, A.Y. Feature selection, L 1 vs. L 2 regularization, and rotational invariance. In Proceedings of the Twenty-First International Conference on Machine Learning, New York, NY, USA, 4–8 July 2004; 2004; p. 78.
83. Srivastava, N.; Hinton, G.; Krizhevsky, A.; Sutskever, I.; Salakhutdinov, R. Dropout: A simple way to prevent neural networks from overfitting. *J. Mach. Learn. Res.* **1929**, *15*, 1929–1958.
84. Petrich, J.; Gobert, C.; Phoha, S.; Nassar, A.R.; Reutzel, E.W. Machine learning for defect detection for PBFAM using high resolution layerwise imaging coupled with post-build CT scans. In Proceedings of the 27th International Solid Freeform Fabrication Symposium, Austin, TX, USA, 7–9 August 2017.
85. Gobert, C.; Reutzel, E.W.; Petrich, J.; Nassar, A.R.; Phoha, S. Application of supervised machine learning for defect detection during metallic powder bed fusion additive manufacturing using high resolution imaging. *Addit. Manuf.* **2018**, *21*, 517–528. [CrossRef]
86. Imani, F.; Gaikwad, A.; Montazeri, M.; Rao, P.; Yang, H.; Reutzel, E. Layerwise in-process quality monitoring in laser powder bed fusion. In Proceedings of the ASME 2018 13th International Manufacturing Science and Engineering Conference, College Station, TX, USA, 18–22 June 2018.
87. Aminzadeh, M.; Kurfess, T.R. Online quality inspection using Bayesian classification in powder-bed additive manufacturing from high-resolution visual camera images. *J. Intell. Manuf.* **2019**, *30*, 2505–2523. [CrossRef]
88. Zhang, Y.; Hong, G.S.; Ye, D.; Zhu, K.; Fuh, J.Y. Extraction and evaluation of melt pool, plume and spatter information for powder-bed fusion AM process monitoring. *Mater. Des.* **2018**, *156*, 458–469. [CrossRef]
89. Baumgartl, H.; Tomas, J.; Buettner, R.; Merkel, M. A deep learning-based model for defect detection in laser-powder bed fusion using in-situ thermographic monitoring. *Prog. Addit. Manuf.* **2020**, 1–9. [CrossRef]
90. Scime, L.; Beuth, J. Using machine learning to identify in-situ melt pool signatures indicative of flaw formation in a laser powder bed fusion additive manufacturing process. *Addit. Manuf.* **2019**, *25*, 151–165. [CrossRef]
91. Scime, L.; Beuth, J. Anomaly detection and classification in a laser powder bed additive manufacturing process using a trained computer vision algorithm. *Addit. Manuf.* **2018**, *19*, 114–126. [CrossRef]

92. Repossini, G.; Laguzza, V.; Grasso, M.; Colosimo, B.M. On the use of spatter signature for in-situ monitoring of Laser Powder Bed Fusion. *Addit. Manuf.* **2017**, *16*, 35–48. [CrossRef]
93. Ye, D.; Hong, G.S.; Zhang, Y.; Zhu, K.; Fuh, J.Y.H. Defect detection in selective laser melting technology by acoustic signals with deep belief networks. *Int. J. Adv. Manuf. Technol.* **2018**, *96*, 2791–2801. [CrossRef]
94. Shevchik, S.A.; Kenel, C.; Leinenbach, C.; Wasmer, K. Acoustic emission for in situ quality monitoring in additive manufacturing using spectral convolutional neural networks. *Addit. Manuf.* **2018**, *21*, 598–604. [CrossRef]
95. Grasso, M.; Demir, A.; Previtali, B.; Colosimo, B. In situ monitoring of selective laser melting of zinc powder via infrared imaging of the process plume. *Robot. Comput. Integr. Manuf.* **2018**, *49*, 229–239. [CrossRef]
96. Okaro, I.A.; Jayasinghe, S.; Sutcliffe, C.; Black, K.; Paoletti, P.; Green, P.L. Automatic fault detection for laser powder-bed fusion using semi-supervised machine learning. *Addit. Manuf.* **2019**, *27*, 42–53. [CrossRef]
97. Yuan, B.; Giera, B.; Guss, G.; McMains, S. *Semi-Supervised Learning for Selective Laser Melting Monitoring via In-Situ Videos*; Lawrence Livermore National Lab.(LLNL): Livermore, CA, USA, 2018.
98. Yao, B.; Imani, F.; Yang, H. Markov decision process for image-guided additive manufacturing. *IEEE Robot. Autom. Lett.* **2018**, *3*, 2792–2798. [CrossRef]
99. Achieve the Highest Possible Quality in Series Production Thanks to LaserCUSING ®, n.d. Available online: www.concept-laser.de (accessed on 30 March 2020).
100. Monitoring for 3D Printing Systems, (n.d.). Available online: https://www.eos.info/software/monitoring-software/meltpool-monitoring (accessed on 30 March 2020).
101. Alberts, D.; Schwarze, D.; Witt, G. In Situ Melt Pool Monitoring and the Correlation to Part Density of Inconel®718 for Quality Assurance in Selective Laser Melting. Available online: https://pdfs.semanticscholar.org/e410/17f193cd01fbb30b97ef140c698ba48e609d.pdf (accessed on 17 June 2020).
102. InfiniAM Spectral, (n.d.). Available online: https://www.renishaw.com/en/infiniam-spectral-42310 (accessed on 30 March 2020).
103. Monitoring TruPrint | TRUMPF, (n.d.). Available online: https://www.trumpf.com/en_US/products/services/services-machines-systems-and-lasers/monitoring-analysis/monitoring-truprint/ (accessed on 30 March 2020).
104. Sigma Labs—Quality Assurance 3D Printing and Additive Manufacturing, (n.d.). Available online: https://sigmalabsinc.com/?page=products (accessed on 30 March 2020).

crystals

MDPI

Article

Finite Element Simulation of Multilayer Electron Beam Melting for the Improvement of Build Quality

Manuela Galati *, Oscar Di Mauro and Luca Iuliano

Department of Management and Production Engineering (DIGEP), Integrated Additive Manufacturing Center (IAM), Politecnico di Torino, 10129 Torino, Italy; oscar.dimauro@polito.it (O.D.M.); luca.iuliano@polito.it (L.I.)

* Correspondence: manuela.galati@polito.it; Tel.: +39-0110904569

Received: 20 April 2020; Accepted: 21 June 2020; Published: 23 June 2020

Abstract: Macroscale modeling plays an essential role in simulating additive manufacturing (AM) processes. However, models at such scales often pay computational time in output accuracy. Therefore, they cannot forecast local quality issues like lack of fusion or surface roughness. For these reasons, this kind of model is never used for process optimization, as it is supposed to work with optimized parameters. In this work, a more accurate but still simple three-dimensional (3D) model is developed to estimate potential faulty process conditions that may cause quality issues or even process failure during the electron beam melting (EBM) process. The model is multilayer, and modeling strategies are developed to have fast and accurate responses. A material state variable allows for the molten material to be represented. That information is used to analyze process quality issues in terms of a lack of fusion and lateral surface roughness. A quiet element approach is implemented to limit the number of elements during the calculation, as well as to simulate the material addition layer by layer. The new material is activated according to a predefined temperature that considers the heat-affected zone. Heat transfer analysis accuracy is comparatively demonstrated with a more accurate literature model. Then, a multilayer simulation validates the model capability in predicting the roughness of a manufactured Ti6Al4V sample. The model capability in predicting a lack of fusion is verified under a critical process condition.

Keywords: roughness; electron beam melting (EBM); surface texture; lack of fusion; part quality; Ti6Al4V; metal additive manufacturing

1. Introduction

Over the years, increased computing power combined with proper software development, has allowed for the implementation of simulation tools to speed up product development [1]. Chief among the numerous advantages offered by simulation is the possibility of investigating complex processes. Especially for new processes, such as additive manufacturing (AM), process simulation looks to be the best tool for processing complex understandings [2]. Electron beam melting (EBM) is an AM powder-bed process for metal components used for the production of end-usable parts in several industrial sectors [2]. During the EBM process, a high energy electron beam melts metallic powders by moving across the powder bed [3]. After the powder distribution, the whole powder bed is preheated. The preheating step establishes a uniform temperature on the powder bed, which is approximately equal to 60% of the melting point of the processed material. Except for a low helium flow (0.002 mbar) [4], the whole process occurs under vacuum to avoid beam deflection and to maintain a hot working environment. Thanks to the high working temperature and vacuum environment, EBM guarantees the possibility of manufacturing parts with high-quality materials such as intermetallic titanium alloys that cannot be processed with other AM or conventional processes [5]. Although the EBM process has great potential, the difficulties in controlling and determining proper process conditions limit the EBM application to niche industrial sectors. Process complexity is a result of numerous phenomena involved in the process [6], including the rapid material phase change [2]. The process conditions and implementation difficulties of process monitoring systems [2] make EBM attractive at the numerical scale. Figure 1 maps the relevant literature to an EBM simulation. Few models have adopted a mesoscopic approach in which particle to particle modeling is conducted [7–11]. The great level of detail included in powder modeling requires significant computational time and resources. Therefore, usually, only two-dimensional (2D) models were implemented. To reduce the computational time, most models in the literature are solved using the finite element (FE) method. In this case, the material is modeled as a continuum. FE models are mainly used to predict thermal distribution (pure thermal or uncoupled) [2,6,12–22]. These models simulated a single track or a single-layer and analyzed the process from a microscopic level (i.e., temperature distribution and melt pool size). Coupled models use a thermal solver coupled with at least one other solver. These models can be grouped into fully coupled thermomechanical [12,22–28], thermo-fluid flow [18,24,29], or weakly coupled [13,14]. A limited category of models analyzed the microscopic aspects of the process, such as the interaction between beam and powder particles [15,30,31], or even microstructural evolution [32]. Owing to the high number of elements that should be involved in the calculation, multilayer simulations have rarely been implemented [33] and instead have been primarily used for predicting residual stresses.

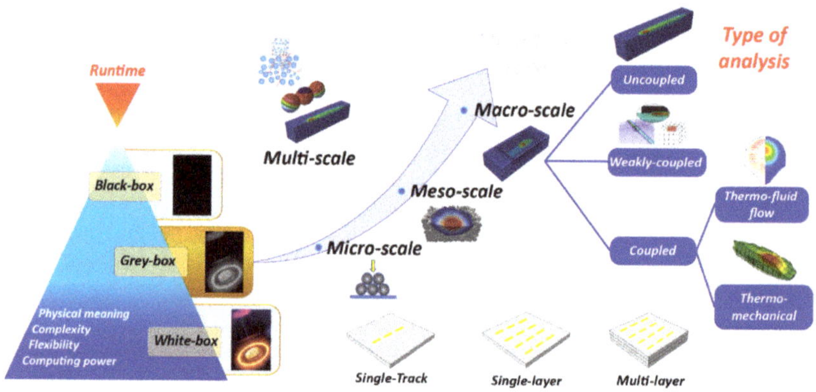

Figure 1. Classification of electron beam melting (EBM) process simulation models.

Except for a few studies [15,16,23], calibrating coefficients were often applied to the heat source model to improve the match between experimental and simulation results. None of the reviewed FE models have explored the possibility of predicting the macroscopic process quality issues, i.e., the lack of fusion or surface roughness profile. As far as surface roughness is concerned, all developed models are empirical and can only estimate the average roughness value of the surface according to its angle with respect to the build platform [34]. S. Shrestha and K. Chou [35] developed a finite volume model based on a free surface approach to predict only the top surface roughness. However, the top surface is usually less critical than the vertical one because it shows better texture and a lower Ra value [34]. Typical values for vertical surface roughness range between 24 and 30 μm [36], while the Ra value of the top surfaces is around 6 μm [34]. Vertical surface roughness is mainly affected by the heat transfer effect between the molten material and the surrounding powder [34,37].

The prediction of macroscopic quality issues such as the lack of fusion and surface roughness profile over a short time is an important topic for simplifying the optimization process. This work presents a new and fast three-dimensional (3D) uncoupled thermal model for the multilayer simulation of the EBM process. The heat source is modeled following the approach developed in previous work [15], in which no calibrating coefficient was used. The multilayer simulation emulates the creation of the part layer by layer and the electron beam control along the hatching path and layer rotation. A material state variable is introduced to represent the material, which is melted during the process. This variable allows the identification of both the surface roughness profile and the lack of fusion. A new procedure is developed to reduce the calculation time, which accounts for a minimum number of elements during the calculation. This procedure is based on the quiet element approach, in which certain elements are activated only when necessary. This approach has been investigated in the literature to activate elements within a predefined volume by the time [38], such as the heat-affected zone (HAZ). However, the HAZ can vary in size and severity depending on the process condition and can be determined only by running experiments. A rough estimation of this quantity involves an incorrect number of elements that participate in the calculation and, therefore, a heat transfer analysis can be inaccurate or slow. To overcome this issue, the proposed procedure automatically considers element activation as a function of a predefined temperature. The activation temperature is determined by observing the HAZ predicted by the microscale model developed in a previous work [15]. The accuracy of the heat transfer analysis carried out with the proposed model, in terms of melt pool size and temperature distribution, is verified against the microscale model [15]. The capability of the proposed model in predicting the lateral roughness is validated by an experimental comparison. Herein, we verify the model response to a critical process parameter condition that causes a lack of fusion.

2. Materials and Methods

2.1. Modeling

Numerical modeling was based on the fundamentals presented in a previous study [15] in which a 3D microscale thermal model for the EBM process was developed. That work presented a new heat source model tailored according to the actual beam impact on the powder bed and new modeling for the powder thermal proprieties. The beam impact was simulated by Monte Carlo simulations. These simulations determined the effective beam diameter D_E that mimics the electron distribution on the powder surface. The energy source model was, therefore, defined as a uniform heat distribution on a circular area in which a coefficient η depends on D_E. This was used to couple the electron beam size with the uniform heat source. The material was modeled as a continuum, and the calculation of actual thermal conductivity of the powder considered both thermal radiation in the pores and heat transfer through the powder necks formed during preheating. The powder bed's specific heat and latent heat of fusion were assumed to be equal to those of the solid bulk material [15,16]. The porosity of the powder material, φ, was considered equal to 0.32, which corresponds with a particle arrangement in body cubic centered (BCC) structure. Because of the thermal properties of the powder differ from the bulk (material after the melting) one, the powder was handled as different material, and during the melting simulation, a material change procedure updated the material properties from powder to bulk according to the achieved temperature. In that and subsequent works [15,16], the model was used to predict and validate the temperature distribution and melt pool size during the melting of a single track. Although the accuracy of this kind of modeling, the modeling of the actual EBM process should consider the melting of the entire layer and the addition of the material layer by layer. With this aim, the heat source developed in Reference [15] was adapted to consider the electron beam path across the powder bed. The heat source was, therefore, calculated according to the following equations:

$$q = \eta UI/S \ (x_1,x_2,x_3) \in \Gamma_i \tag{1}$$

$$q = 0 \ (x_1,x_2,x_3) \in (D - \Gamma_i) \tag{2}$$

where η is calculated according to the procedure [15], D is the whole simulation domain, and Γi is the domain of the flux application. The shape of Γi is circular and corresponds to the cross-section of the beam diameter. x_1 and x_2 are the coordinates of the nodes in the plane of the layer, while x_3 is the coordinate along the build direction, which is the axis normal to the building platform (Figure 2). If n_{layer} is the total number of layers to be simulated, x_3 varies layer after layer according to Equation (3).

$$x_3 = -(n_{layer} + n - 1)\Delta s \tag{3}$$

The index n is the index of actual layer and Δs is the layer thickness and thus an integer variable, which is calculated (Equation (4)) as the ratio between the total time increment, t, and time to scan a single layer, t_{layer}.

$$n = int(t/t_{layer}) \tag{4}$$

The beam movement over the x_1x_2 plane is governed by Equation (5) that describes the selection of the nodes that belong to the domain Γ_i in the x_1x_2 plane.

$$(x_1 - x_{1,inc})^2 + (x_2 - x_{2,inc})^2 \leq (\varnothing/2)^2 \tag{5}$$

where Ø is the beam diameter and $x_{1,inc}$ and $x_{2,inc}$ are the incremental coordinates from the first position of the beam on the plane (x_{1c0}, x_{2c0}) and the last point of the scanning path. The beam is, therefore, moved point by point along a predefined scanning path. In this work, a standard single direction hatching strategy (Figure 2) was implemented. According to this strategy, the beam is moved with a constant speed, v, across the powder bed along parallel lines. The line offset parameter (LO) establishes the distance between the centerline of two adjacent lines, and all hatch melt lines are in the same direction. At each layer, the scanning path is rotated by 90 degrees clockwise. Therefore, $x_{1,inc}$ and $x_{2,inc}$ are calculated as follows:

$$x_{1,inc} = x_{1c0} + a_{12}[v(t - nt_{layer}) - mt_{track}] + a_{11}mLO \tag{6}$$

$$x_{2,inc} = x_{2c0} + a_{22}[v(t - nt_{layer}) - mt_{track}] + a_{21}mLO \tag{7}$$

where a_{ii} are the coefficients of an antisymmetric matrix that considered the single direction hatching and the scanning path rotation (Figure 2). The a_{ii} elements were calculated as follows:

$$a_{11} = a_{22} = -\sin(n\pi/2), \tag{8}$$

$$a_{12} = -a_{21} = \cos(n\pi/2). \tag{9}$$

In Equations (6) and (7), m is the line index and represents the actual line to be melted. The line index, m, was calculated from the time for the single line, t_{tracks}, as follows:

$$m = int[(t - nt_{layer})/t_{track}] \tag{10}$$

Since the simulation here considered the melting of a square of side l_{tot}, t_{track} was calculated as follows:

$$t_{track} = (l_{tot} - \varnothing)/v \tag{11}$$

The addition of the material layer by layer was emulated using a new element activation procedure based on the quiet element approach [33]. This method is a numerical approach used in FE simulation for processes that need to account for addition of material during the process. All elements of the model must be modeled from the beginning. Therefore, all elements within the workspace are assembled into the global thermal conductivity matrix, but the elements yet to be deposited are assigned 'void' material properties. Practically, the quiet elements are elements in which the thermal properties are adequately scaled and do not affect the calculation. The domain of analysis is progressively increased according to specific conditions. In this work, the quiet element approach was used not only to simulate the layer addition but also to keep the number of elements involved in the calculation into the current melted layer as low as possible. In this sense, the quiet elements confine the heat exchange only in the HAZ. This strategy is expected to increase the computational speed with respect to the literature models that consider the entire layer at the beginning of the simulation. This method uses a constant mesh and is simple. However, incorrect selection of elements to be eventually activated can lead to ill heat transfer. For this reason, a new criterion was implemented that aims to determine the lowest number of elements belonging to the HAZ without decreasing the accuracy of the heat transfer. Practically, this corresponds to determining the minimum temperature that refers to the location of the HAZ boundary at which the element must be "activated". This temperature is called "activation temperature". Element activation means changing the element material properties to the real ones. This methodology is called thermal activation because it depends exclusively on heat transfer. Since the nodal temperatures of a quiet element are different from those of an active element, body heat flux is numerically generated in some integration points to compensate for the temperature difference [39]. In this way, when the heat source moves, heat transfer can occur. This propagation generates a temperature field in the nodes of the quiet elements, which are constantly compared with the predefined activation temperature.

The determination of HAZ during the melting involves temperature monitoring and, therefore, experimental trials. To overcome these issues, the microscale model [15,16] was used to determine the HAZ, and thus the activation temperature. Firstly, few melting tracks are simulated using the microscale model. During the simulation, the minimum temperature that corresponds to the boundary of HAZ is identified. This temperature is set as activation temperature in the macroscale model, and the same simulation is performed. The results from the two models are compared in terms of temperature distribution and melt pool size. The macroscale model is then calibrated by changing the activation temperature iteratively until a good match between the two models is observed. With this approach, a trade-off between runtime, numerical convergence, and simulation accuracy is guaranteed. Several simulations at different process conditions showed that the activation temperature is slightly higher than the preheating temperature used in the model.

To further reduce the running time, material change was not considered. Therefore, the material properties were modeled directly as bulk material within the solid part are simulated. However, the material surrounding the external surfaces was modeled as powder material in order to account for the heat transfer between the melted and surrounding materials.

For each node of the model, a material state variable ID was defined to indicate the material's temperature history. The material state was defined as "unmelted" (ID = 0) before the melting point was reached and as "melted" (ID = 1) after the melting point was reached. At the beginning of the process, all elements had an ID equal to zero. This variable, therefore, identified the material melted during the EBM process simulation from the one not melted. The melted material state (ID = 1) was used for both the liquid state and for all solid material in the model. The elements for which the ID was still equal to zero at the end of the process were identified as having a lack of fusion areas.

2.2. Model Implementation

The thermal model was implemented in Abaqus/Standard using user code subroutines for the heat source motion, element activation, and molten material representation. The whole domain is rather simple and consists of a solid substrate on which a set of layers are laid. To reduce the calculation time, the analysis domain used was a small portion of a specimen produced by EBM process with an external lateral surface in contact with a certain volume of loose powder. This powder volume was included to consider the heat exchange between the melted part and the surrounding area. The whole domain was divided into four subdomains (Figure 2): the bulk substrate, the powder substrate, the bulk quiet domain, and the quiet powder domain. The bulk substrate represented material already processed and not simulated in the analysis. The powder substrate was adjacent to the solid substrate. Therefore, a vertical surface of the bulk domain was numerically coupled with one surface of the powder domain. The other surfaces were considered adiabatic. The quiet bulk domain represented the domain scanned by the electron beam. It contained quiet elements that were activated once they achieved the activation temperature. The quiet powder domain represented a portion of material adjacent to the quiet bulk domain that was not scanned. This domain contained quiet elements that were automatically activated at the beginning of the layer and simulated heat transfer from the electron beam track and the surrounding powder. The quiet powder and bulk domains share a lateral surface. The other surfaces were considered adiabatic.

The mesh included DC3D8 (heat transfer brick) elements with a size equal to 0.05 mm in the quiet bulk and powder domains. The rest of the domain was discretized with a progressive coarsening mesh of up to 0.5 mm with DC3D4 elements (heat transfer tetrahedron).

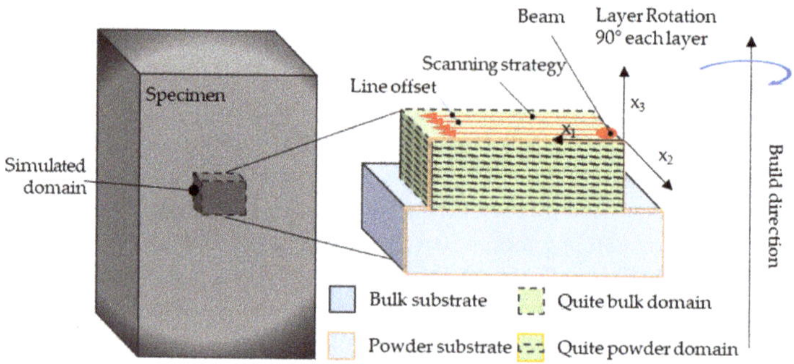

Figure 2. Finite element (FE) model configuration.

The overall domain was subjected to a predefined field corresponding to the preheating temperature. To simulate the preheating step, the elements activated during the simulation had a preheating temperature.

According to the EBM melting strategy, the heat source swiped across the layer using a hatching mode. A specific user code, DFLUX, controlled the electron beam movement according to a user code that implemented the Equations from (1) to (11). Initially, the heat source moved at a constant speed along a certain axis. At each time increment, the code identified the nodes with the domain Γ_i and applied the heat flux according to Equations (1) and (2). After the scanning of a track, the heat source was repositioned at the origin of the previously scanned line. Then, it was translated perpendicularly to the previously scanned line and was of a quantity equal to LO. Hence, a new track was melted. The process was repeated until the layer was completed. Since only a portion of the whole cube was simulated, the code also considered the idle times due to the non-simulated portion of the layer.

To consider the melting strategy rotation, the scan lines were tilted on every layer by 90 degrees (Figure 2). Therefore, the code considered the heat source path rotation and its translation along the build direction. Both the path rotation and the translation were controlled by the time and the scanning speed.

The addition of the material during the simulation was implemented using the UEPACTIVATION VOL subroutine. This subroutine allows for the element activation control using rules predefined by the user. According to the thermal activation described in the previous paragraph, the elements are activated if the nodal temperature (T) is higher than the activation temperature (T_{act}). Owing to the Abaqus limitation, at the beginning of each layer, a few elements must be activated to trigger the subsequent thermal activation. Therefore, a predefined set of elements corresponding to the beam cross-section was activated by default at the beginning of each layer. This initial set was concentrically positioned at the center of the first beam spot. During subsequent increments, the elements were activated via the achieved temperature. Because of the heat propagation, it was possible that subsequent melting would determine the activation of elements belonging to the underlying layer if the temperature was higher than the activation temperature.

At the beginning of the simulation, the ID value at each node was equal to zero. At each increment, the ID was updated using the USDFLD subroutine. This user code identified and collected the nodes that had achieved the solidus temperature ($T_{Solidus}$), while the temperature at the material point for each increment (T) was read using the GETVRM subroutine. Practically, all ID values were read at each node at the end of each increment, and the ID values equal to zero were updated (from 0 to 1) if the node achieved the melting point.

Figure 3 summarizes the FE whole structure implementation. Appendix B reports the source code of DFLUX and UEPACTIVATIONVOL subroutines as implemented in Abaqus.

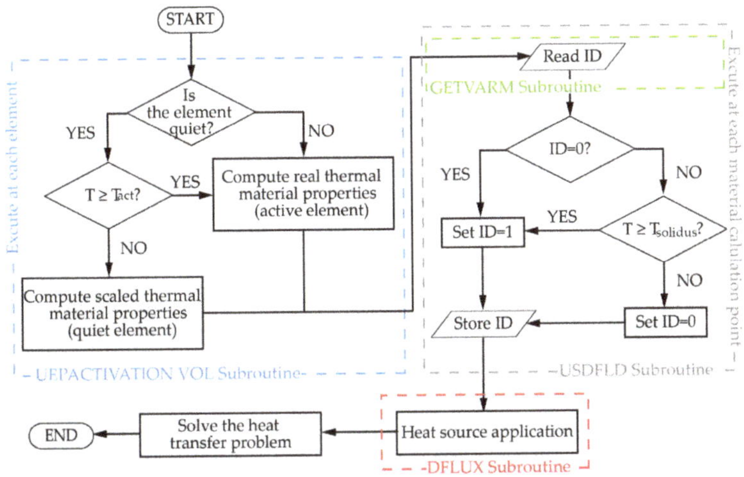

Figure 3. FE model structure and the subroutine for the material addition (UEPACATIVATION VOL), the material state variable for the melted material (USDFLD), and heat flux application (DFLUX) implementations.

2.3. Model Validation

As mentioned above, the proposed model needs proper calibration before it can be run. The calibration is performed using the microscale model [15], named the reference model. Practically, the activation temperature in the macroscale model is iteratively changed until similar heat transfer conditions between the models are obtained.

First, to demonstrate the reliability of the heat transfer obtained via the proposed model after the calibration, a preliminary validation shows the comparison between the proposed and the reference models in terms of temperature distribution and melt pool size. This analysis considered the melting of a portion of a single layer. Then, further validation evaluates the ability of the model to predict the lateral surface roughness. In this case, a multilayer simulation was performed and compared with the experimental data.

In order to give a measure of the computational time, all simulations were performed on a personal computer with the following specifications: Intel® Core ™ i7-8700K 3.70 GHz processor, 32 GB RAM, and 1 TB SSD. A six-core parallelization was used according to the computational resources used to perform the analysis.

2.3.1. Comparison with the Reference Model: Single Layer Validation

This validation was carried out by simulating the melting of a single layer, 30 mm × 30 mm, of Ti-48Al-2Cr-2Nb alloy. The layer was 0.090 mm thick. The material properties for the bulk material were extracted from Reference [40], while the powder proprieties were calculated from the bulk one according to Reference [15]. The material properties implemented in the model were reported as Supplementary Materials. The simulation by the proposed and reference models consisted of the melting of six unidirectional parallel scan tracks. Table 1 lists the process parameters used for both simulations. The comparison considered both the melt pool size and temperature distribution.

Table 1. Simulation parameters for both analyses.

Parameters	
Beam diameter, D [mm]	0.633
Line offset [mm]	0.200
Scan speed, v [mm/s]	2800
η	0.50
Heat flux [W/mm^2]	790.140
Room temperature, T_r [K]	1323
Preheat temperature, T_{pr} [K]	1323
Diameter of the starting zone D_{sz} [mm]	0.700

2.3.2. Multilayer Experimental Validation

The experiment consisted of producing a 10 mm × 10 mm × 20 mm-sized parallelepiped sample made on Ti6Al4V. The sample was produced using an Arcam A2X, standard Arcam powder (average particle size equal to 75 µm) with a layer thickness equal to 0.050 mm. The process parameters are listed in Table 2. According to the standard melting strategy, the melting strategy consisted of unidirectional parallel lines with a clockwise rotation of 90 degrees every layer (Figure 2). After the production, the sample was cleaned by loose powder. The 2D surface roughness profiles of the lateral surface samples were acquired using an RTP-80 profilometer, equipped with a TL90 drive unit. The cut-off length and sampling length were chosen according to ISO 4288:1997. For each lateral surface of the sample, the roughness profile was measured along the build direction. Additionally, for each surface, three profiles were collected: one on the center of the specimen and the other two equally spaced from the first one and from the sample's edges. The collected profiles for each surface are reported in Appendix A, while all collected data and roughness descriptors can be found in the database [41]. The average roughness (Ra) and root mean square (Rq) values were equal to 28 µm and 33 µm, respectively.

Table 2. Process parameters for the experiment and simulation.

Parameters	
Beam diameter, D [mm]	0.780
Line offset [mm]	0.200
Scan speed, v [mm/s]	471.69
η	0.51
Heat flux [W/mm^2]	347.920
Room temperature, T_r [K]	923
Preheat temperature, T_{pr} [K]	923

The numerical simulation by the proposed model emulated the experimental setup. The material properties for Ti6Al4V were extracted from a prior study [15]. The layer was fixed to be equal to 0.050. Figure 2 shows the portion of the simulated specimen. The simulated portion of the cube was 2.2 mm × 2 mm × 1.5 mm. As described above, the analysis domain consisted of four subdomains. The quiet bulk domain contained 10 layers (soft green with black dotted edges in Figure 2) that were added over a solid substrate (soft blue with black edges in Figure 2). The powder substrate (soft blue with orange edges) shared a lateral surface with the solid substrate, while the quiet powder domain (soft green with orange dotted edges in Figure 2) was located over the powder substrate and adjacent to the quiet bulk domain. According to the proposed procedure, at the beginning of each layer, the initial activated area was cylindrical and concentric to the initial position of the beam. The height of the cylinder was set to be equal to the layer thickness, and its diameter was equal to the beam diameter. The video of the simulation after the model calibration is provided as Supplementary Materials.

3. Results and Discussion

3.1. Model Validation

3.1.1. Single-Layer Simulation Validation

As mentioned above, the reliability of the proposed model response was compared with the reference one in terms of melt pool dimensions and maximum temperature (T_{max}). The length (l), the width (w), and the depth (d) of the melt pool were measured. The activation temperature was fine-tuned to a convergence in the heat transfer between the two models. The model was considered calibrated when small deviations were observed between the two models. For an activation temperature equal to 1700 K, the deviations from the reference model are reported in Table 3. This deviation can be attributed to the different material properties attributed to the layer. The reference model, in fact, considered the simulation of the entire layer with the powder thermal material properties that were then updated from powder to bulk according to the achieved temperature during the process. In the proposed model, the portion of the layer to be melted was modeled directly as bulk material. Because of that, deviations were more significant in the first track. In this case, the beam should melt a track surrounded completely by material with low conductivity (powder), while in the proposed model, the powder was modeled only on one side of the beam track. In subsequent passages, the beam overlapped with a portion of the already solidified bulk material, and the deviation between the two models was, therefore, lower, as shown in the third scanning line (Table 3). The difference in the melt pool depth remained, implying that there was an overestimation of the heat penetration in the proposed model. However, the penetration depth was also affected by uncertainty due to the substrate temperature after melting, which could not have been forecasted with a single-layer simulation.

Table 3. Single-layer validation: the deviations between the reference model against the proposed model in terms of the melt pool dimension and maximum temperature.

	T_{max} [%]	l [%]	w [%]	d [%]
Track 1	−11%	32	−7	44
Track 3	−0.14	10	−0.1	44

Overall, the numerical results obtained adopting the new modeling approach can be considered to be consistent with the reference model.

Additionally, using the same computational resources, the total simulation with the reference model took around 8 CPU hours, while the same simulation with the proposed model after calibration took around 5 h. Therefore, the proposed approach to the EBM modeling with proper calibration of the activation temperature allows to obtain an accurate enough heat transfer simulation and a significant runtime saving of about 40% with respect to the reference model.

3.1.2. Lateral Surface Roughness Prediction

According to the proposed approach, firstly, few melting lines were simulated using both the reference and the proposed models. Then, the proposed model was calibrated by changing the activation temperature iteratively until a heat transfer convergence between the two models. For an activation temperature equal to 1000 K, the obtained numerical results from the calibrated model showed a small deviation in comparison with the corresponding reference model measurements. For this reason, the activation temperature was set equal to 1000 K.

To be numerically consistent with the experiment, the surface roughness was described by the surface profile between the quiet bulk and powder domains, considering the melted material.

This information was extracted by the material state variable ID. As mentioned above, the ID was equal to 1 if the node achieved the solidus temperature, otherwise, it remained equal to zero. Graphically, this result can be represented by a colorful map in which the elements in red were an ID equal to 1 and those in blue were an ID equal to zero. Figure 4 shows the experimental 2D roughness profile (blue line) overlapped with the numerical one. The numerical profile was obtained using a cross-section of the model. The surface roughness profile of the sample (blue line) was characterized by peaks and valleys, which were repeated with a certain frequency. This frequency seems to be linked to the rotation layer and, therefore, to a distance equal to four layers. This frequency was also observed in the numerical results. The peaks of the numerical simulation appeared to be representative of the experimental ones, while the numerical valleys appeared to be more profound. This could be explained by the presence of powder particles attached to the experimental surface because of the high sintering level [42,43]. These particles could not be removed during cleaning because they were partially melted into the surface. Since the simulation approach used the FE method and a continuum domain, the presence of these particles was not detected.

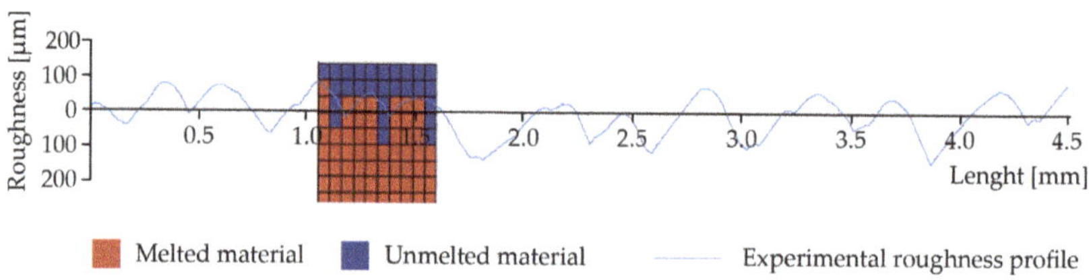

Figure 4. Comparison between the numerical and experimental roughness profiles.

The differences of the wave width between the experimental and the numerical profiles can be explained by the effective layer thickness [12], which was around 2 or 3 times larger than the lowering of the build platform [23] and was not considered in this simulation.

A certain effect may also be caused by the material change, which was not considered in the simulation and may increase the heat transfer toward the powder domain because of the local thermal conductivity change.

Overall, the numerical simulation appeared to represent the actual surface profile well and established the capability of the model to give information about the surface texture resulting from the EBM process.

Figure 5 shows a portion of the 3D profile of the lateral surface obtained using numerical simulation. This profile was obtained by manipulating the numerical output visualization to show the boundary surface between the melted and unmelted material. Figure 5 detects the peaks and valleys along the surfaces previously observed in both 2D experimental and numerical profiles.

To estimate the average roughness (Ra) and the root mean square (Rq) of the surface and to increase the sampling length in the model, a longer simulation was performed up to 17 layers, which corresponds to the height of the simulated sample equal to 0.85 mm. From this simulation, Ra and Rq resulted in being equal to 35 µm and 42 µm, respectively (see Appendix A for details). The obtained values are comparable with the experimental one (Ra = 28 µm and Rq = 33 µm). Therefore, the results obtained from the model are believed to serve as good indicators of the lateral surface profiles. However, as will be discussed in the Conclusion, a more detailed investigation of both experimental and simulated melting is planned.

3.2. *Lack of Fusion in EBM Process*

Given the results presented in Section 3.1.2, further analysis was carried out to evaluate the model response in predicting the lack of fusion under a critical process condition. Lack of fusion occurs easily if the EBM conditions are not appropriate. For instance, an increased line offset toward beam diameter results in lower energy densities, which causes a lack of fusion [44]. With this scope, the line offset of the previous multilayer simulation was increased up to 0.800 mm, which was approximately equal to the beam diameter (Table 2). The other parameters remained constant, including the activation temperature.

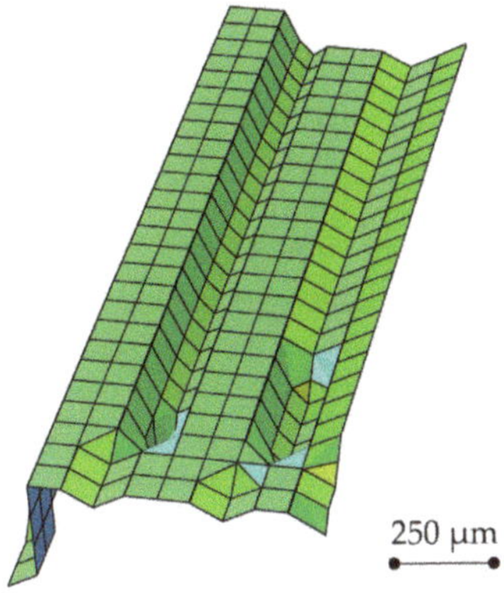

Figure 5. 3D numerical surface profile: peaks and valleys can be observed.

Numerically, the lack of fusion was represented by those elements named "unmelted material" and, therefore, did not achieve the melting point during the simulation. The ID variable gathered the state of the element. As before, within the quiet bulk domain, the unmelted elements (ID = 0) were plotted in blue (Figure 6), while the melted elements (ID = 1) were plotted in translucent. As can been seen, Figure 6 shows the presence of numerous lack-of-fusion areas entrapped in the solid material. The locations of these unmelted areas look to be dependent on the layer rotation strategy. The simulation, therefore, identified this process condition as faulty. In fact, larger line offset values that did not allow for the full melting of the material, causing the lack of fusion appearance, voids, and porosities, thus inhibiting the continuation process [18].

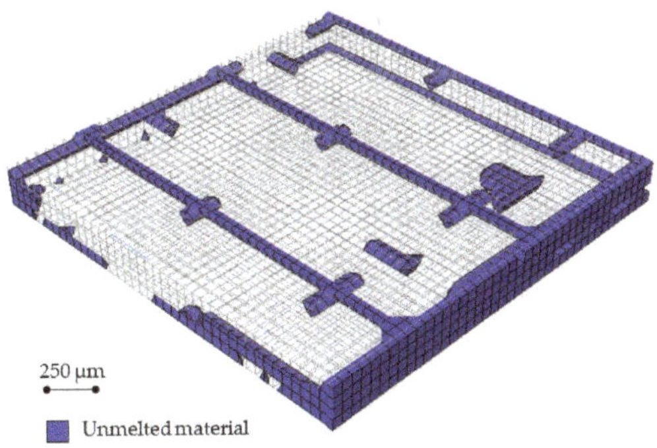

Figure 6. Simulation of a process condition that causes the appearance of a lack of fusion. The translucent elements indicate the material that achieved the melting point during the process (ID = 1), while blue ones indicate the material that has never achieved the melting point (ID = 0). After the cooling phase, the unmelted material during the process represents a lack-of-fusion defect in the final part.

3.3. *Runtime Considerations*

The total duration of the 10-layer simulation was around 20 h. The process simulation with a larger line offset took around 2 h.

A convergence analysis was carried out on mesh size during the activation temperature calibration. The mesh size was finer than 0.050 mm and significantly increased the runtime without increasing the output accuracy. Lower activation temperatures lead to the activation of a higher number of elements, therefore, worsening the runtime. Higher temperatures caused the non-convergence of the model because of the incorrect heat transfer.

4. Conclusions

The literature review of multilayer simulations for the EBM process and more in general for AM processes is poor and did not allow the prediction of the local quality. In this work, an accurate but simple 3D multilayer model was developed for the optimization of process parameters. The aim of the model was to rapidly identify a set of potentially faulty process conditions. To speed up the runtime, we developed a new tailor-made procedure. This procedure was based on the progressive element activation under a specific temperature, which was used not only to simulate the layer by layer production but also to limit the number of elements that participated in the calculation during the single-layer melting. The activation temperature was determined by accurately analyzing the HAZ simulation using a microscale model that predicted the temperature distribution and the actual melt pool. The activation temperature was calibrated, and a small difference in heat transfer was observed between the two models.

To identify the quality inherent in the lack of fusion, our methodology included the definition of a material state variable (ID), which allowed us to distinguish between melted and unmelted material. The graphical visualization of ID evaluated the roughness profile of a sample surface along a build direction, as well as the lack of fusion appearance. For validation purposes, the comparison between the microscale model used for the activation temperature calibration and the proposed one was shown. Successively, the model's ability to predict surface texture was verified against experimental results. The model response was tested under a critical condition that, in real production, generated a process failure. The right activation temperature led to a small deviation between the microscale and the proposed models in terms of heat transfer with a savings time of about 40%. The model showed good agreement with the experimental and numerical measurements in forecasting the 2D surface profile and the roughness descriptors (Ra and Rq). The little deviation between the experimental and the numerical demonstrates that the proposed approach will be an efficient model for simulating the roughness profile. However, a more detailed numerical analysis will be carried out for the verification of the model capability in roughness prediction at different process conditions. Additionally, further work will consider implementing the effect on the powder layer due to account both for thermal expansion of powder particles and for porosity reduction of the powder bed during the heating [23].

The model also demonstrated the ability to predict a lack of fusion associated with thermal effects due to, for example, a large line offset. In such a case, with respect to the experimental trial, the simulation allowed for hugely significant time-saving process optimization. From this point of view, the presented results demonstrated that the proposed approach could be easily applied to the process development stage in order to reject faulty process conditions. From an industrial point of view, this approach can save time and resources with respect to the experimental routine. Further model validations and developments with the aim of simulating complex parts may help in the production of defect-free parts. Additionally, an accurate prediction of surface profile roughness may allow for the identification of minimum allowances, which are needed to add for the subsequent finishing operations. Moreover, there is a need for microcrack identification, which can jeopardize a product's fatigue limits.

Supplementary Materials: The following are available online at http://www.mdpi.com/2073-4352/10/6/532/s1, **S1**: Video: Multilayer process simulation of the EBM process for Ti-6Al-4V: the size of the simulated domain of material addition is 2.2 mm × 2 mm × 1.5 mm. The domain contains 10 layers with a layer thickness equal to 0.050 mm. The process parameters are: beam diameter = 0.780 mm; Line offset = 0.200 mm; scan speed = 471.69; Heat flux = 347.920 W/mm^2; Preheat temperature = 923 K. The movie shows the temperature distribution evolution over the time (on the right) and the corresponding melted material (on the left). **S2**: Thermal material properties for Ti-48Al-2Cr-2Nb implemented in the simulation.

Author Contributions: Conceptualization, M.G.; data curation, M.G. and O.D.M.; investigation, M.G. and O.D.M.; methodology, M.G.; resources, L.I.; software, M.G. and O.D.M.; supervision, L.I.; validation, M.G. and O.D.M.; visualization, M.G. and O.D.M.; writing—original draft, O.D.M.; writing—review and editing, M.G. All authors have read and agreed to the published version of the manuscript.

Funding: This research received no external funding.

Conflicts of Interest: The authors declare no conflict of interest.

Appendix A. Experimental and Numerical Roughness Profile and Ra and Rq Calculation

Figure A1 depicts all surface roughness collected during the experiments. For each surface of the sample, the roughness profiles have been acquired along the build direction and replicated three times. The first profile was acquired along the surface center, while the other two profiles were equally spaced from the first one and the edges of the sample. These measurements were collected and indexed as follows: the first number (from 1 to 4) indicates the index of the surface; the second number (from 1 to 3) specifies the measure replica. The surface roughness profiles of the sample were acquired using an RTP-80 profilometer, equipped with a TL90 drive unit. The cut-off length and sampling length were chosen according to ISO 4288:1997.

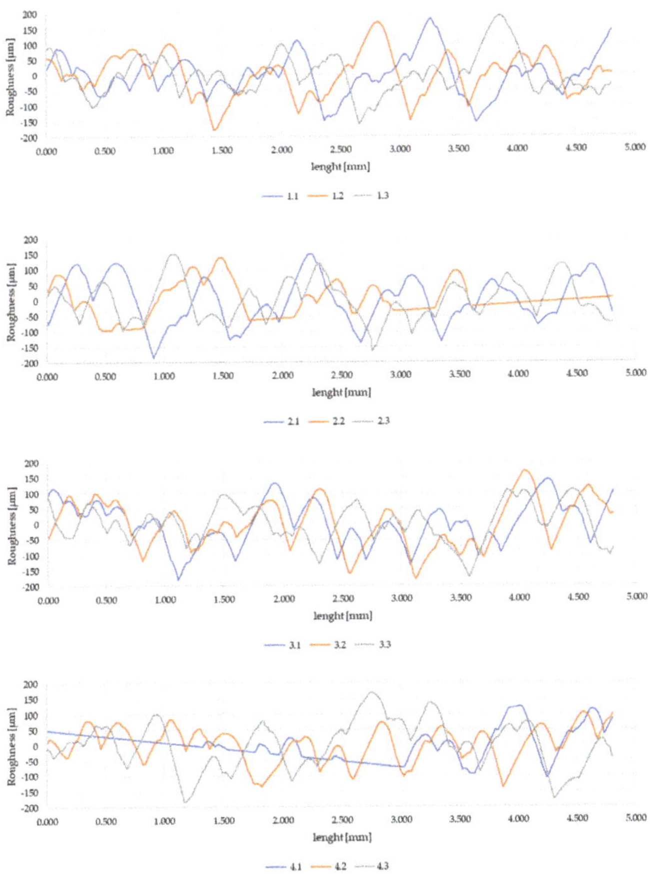

Figure A1. Surface roughness profiles from each surface. The measurements have been replicated three times. The first number (from 1 to 4) indicates the index of the surface; the second number (from 1 to 3) specifies the measure replica.

For each measurement, the parameters Ra and Rq and the surface profile were also collected. The average value of the Ra and Rq was equal to 28 μm and 33 μm, respectively. All collected measurements and roughness descriptors can be found in the database [41].

A portion of the samples has been simulated numerically according to the proposed methodology. In order to have a significant and reliable number of roughness values to calculate the predicted Ra and Rq values, the addition of 17 layers over a solid substrate has been simulated. According to the methodology presented in the paper, the boundary between the unmelted and the melted materials has been extracted from the simulation results by using the ID variable (Figure A2a). According to the standard [45], Ra is calculated, over the evaluation length, as the arithmetical mean of the absolute values of the profile heights (y_i) from the mean line of the effective profile. The mean line is the line having the shape of the geometrical profile and dividing the effective profile (Figure A2) so that within the sampling length, the sum of the squares of the distances between the effective profile points and the mean line is minimum. Analogously, Rq is the root mean square average of the profile heights over the evaluation length. From the numerical model, Ra and Rq resulted in being equal to 35 μm and 42 μm, respectively.

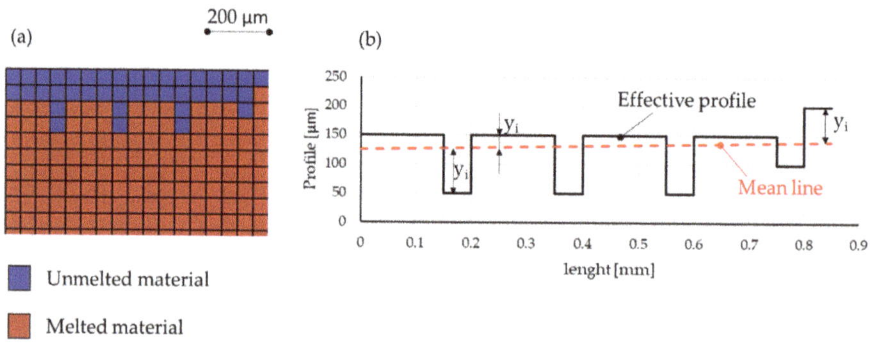

Figure A2. Numerical measurements of the surface roughness: (**a**) cross-section of the model that shows the ID state variable map (**b**) Numerical profile and its mean line for Ra calculation.

Appendix B. Source Code Listing of DFLUX and UEPACTIVATIONVOL Subroutines as Implemented in Abaqus

Appendix B lists the user code as written in Abaqus to implement the modeling as discussed in the main paper. The subroutines are in FORTRAN language.

```
SUBROUTINE DFLUX(FLUX,SOL,KSTEP,KINC,TIME,NOEL,NPT,COORDS,
JLTYP,TEMP,PRESS,SNAME)
INCLUDE 'ABA_PARAM.INC'
DIMENSION FLUX(2), TIME(2), COORDS(3)
CHARACTER*80 SNAME
c
ltot = 10 !scan track lenght
D = 0.78 !beam diameter
V = 471.69 !beam speed
Q = 347920 !heat flux magnitude
LO = 0.2 !line offset
LT = 0.05 !layer thickness
n_layer = 10 !number of layers
c
x1 = COORDS(3) !spatial coordinate x1
```

```
x2 = COORDS(1) !spatial coordinate x2
  x3 = COORDS(2) !spatial coordinate x3
  t = TIME(1) !total time
c
x1c0 = 0 !coordinates of the first position of the beam
  x2c0 = −D/2 !coordinates of the first position of the beam
c
t_track = (ltot-D)/v !time to perform a single line
t_layer = ltot/LO*t_track !time to scan a single layer, tlayer
c
n = int(t/t_layer) !layer index
m = int((t-n*t_layer)/t_track) !line index
c
a11 = −sin(n*(pi/2)) !matrix coefficients
  a22 = −sin(n*(pi/2))
  a12 = cos(n*(pi/2))
  a21 = −cos(n*(pi/2))
c
x1_inc = x1c0 + a12*(v*(t-n*t_layer)-m*t_track) + a11*m*LO !incremental coordinates from x1c0 and the last point of the scanning path
  x2_inc = x2c0 + a22*(v*(t-n*t-layer)-m*t_track) + a21*m*LO !incremental coordinates from x2c0 and the last point of the scanning path
c
if(x3.EQ.-(n_layer + n − 1)*LT)then !check the x3
if ((x1-x1_inc)**2 + (x2−x2_inc)**2.LE.(D/2)**2)then
  FLUX(1) = q !Heat source application. Since only a portion of the whole cube was simulated,
  end if !was considered the idle times due to the non-simulated portion of the layer
  end if
c
RETURN
END
c
subroutine UEPACTIVATIONVOL(lFlags, epaName, noel, nElemNodes, iElemNodes, mcrd, coordNodes, uNodes, kstep, kinc, time, dtime, temp, npredef, predef, nsvars, svars, sol, solinc, volFract, nVolumeAddEvents, volFractAdded, csiAdded)
include 'aba_param.inc'
character*80 epaName
c
ltot = 10 !scan track lenght
D = 0.78 !beam diameter
V = 471.69 !beam speed
LO = 0.2 !line offset
  LT = 0.05 !layer thickness
n_layer = 10 !number of layers
Tact = 1000 !activation temperature
c At the beginning of the simulation the status of all elements must be set as "quiet"
volFractAdded(1) = 0.d0
t = TIME(1) !total time
x1 = coordNodes(3,1) !spatial coordinate x1
x2 = coordNodes(1,1) !spatial coordinate x2
```

```
x3 = coordNodes(2,1) !spatial coordinate x3
c
x1c0 = 0 !coordinates of the first position of the beam
x2c0 = −R !coordinates of the first position of the beam
  n = int(t/t_layer) !layer index
c Activation of the initial set of elements
if(x3.EQ.-(n_layer+n-1)*LT)then
if ((x1-x1c0)**2+(x2-x2c0)**2.LE.(D/2)**2)then
volFractAdded(1) = 1.d0
end if
end if
c Thermal activation
if (x3.LE.-(n_layer+n-1)*LT)then
if(sol(1).GE. Tact) then
end if
end if
c
return
end
```

References

1. Dilberoglu, U.M.; Gharehpapagh, B.; Yaman, U.; Dolen, M. The role of additive manufacturing in the era of industry 4.0. *Procedia Manuf.* **2017**, *11*, 545–554. [CrossRef]
2. Galati, M.; Iuliano, L. A literature review of powder-based electron beam melting focusing on numerical simulations. *Addit. Manuf.* **2018**, *19*, 1–20. [CrossRef]
3. Gaytan, S.M.; Murr, L.E.; Medina, F.; Martinez, E.; Lopez, M.I.; Wicker, R.B. Advanced metal powder based manufacturing of complex components by electron beam melting. *Mater. Technol.* **2009**, *24*, 180–190. [CrossRef]
4. Markl, M.; Lodes, M.; Franke, M.; Körner, C. Additive manufacturing using selective electron beam melting. *Weld. Cut.* **2017**, *16*, 177–184.
5. Baudana, G.; Biamino, S.; Ugues, D.; Lombardi, M.; Fino, P.; Pavese, M.; Badini, C. Titanium aluminides for aerospace and automotive applications processed by Electron Beam Melting: Contribution of Politecnico di Torino. *Met. Powder Rep.* **2016**, *71*, 193–199. [CrossRef]
6. Sigl, M.; Lutzmann, S.; Zaeh, M.F. Transient physical effects in electron beam sintering. In Proceedings of the 17th Solid Freeform Fabrication Symposium SFF 2006, Austin, TX, USA, 14–16 August 2006; pp. 464–477.
7. Attar, E. Simulation of Selective Electron Beam Melting Processes. Ph.D. Thesis, Friedrich-Alexander-Universität Erlangen-Nürnberg (FAU), Erlangen, Alemanha, 2011.
8. Ammer, R.; Markl, M.; Ljungblad, U.; Körner, C.; Rüde, U. Simulating fast electron beam melting with a parallel thermal free surface lattice Boltzmann method. *Comput. Math. Appl.* **2014**, *67*, 318–330. [CrossRef]
9. Ammer, R.; Rüde, U.; Markl, M.; Körner, C. Modeling of Thermodynamic Phenomena with Lattice Boltzmann Method for Additive Manufacturing Processes. In Proceedings of the 23rd International Conference on Discrete Simulation of Fluid Dynamics (DSFD), Paris, France, 28 July–1 August 2014.
10. Klassen, A.; Forster, V.E.; Körner, C. A multi-component evaporation model for beam melting processes. *Model. Simul. Mater. Sci. Eng.* **2017**, *25*, 025003. [CrossRef]
11. Yan, W.; Qian, Y.; Ge, W.; Lin, S.; Liu, W.K.; Lin, F.; Wagner, G.J. Meso-scale modeling of multiple-layer fabrication process in selective electron beam melting: Inter-layer/track voids formation. *Mater. Des.* **2018**, *141*, 210–219. [CrossRef]
12. Riedlbauer, D.; Scharowsky, T.; Singer, R.F.; Steinmann, P.; Körner, C.; Mergheim, J. Macroscopic simulation and experimental measurement of melt pool characteristics in selective electron beam melting of Ti-6Al-4V. *Int. J. Adv. Manuf. Technol.* **2016**, *88*, 1–9. [CrossRef]

13. Koepf, J.A.; Soldner, D.; Ramsperger, M.; Mergheim, J.; Markl, M.; Körner, C. Numerical microstructure prediction by a coupled finite element cellular automaton model for selective electron beam melting. *Comput. Mater. Sci.* **2019**, *162*, 148–155. [CrossRef]
14. Soldner, D.; Mergheim, J. Thermal modelling of selective beam melting processes using heterogeneous time step sizes. *Comput. Math. Appl.* **2019**, *78*, 2183–2196. [CrossRef]
15. Galati, M.; Iuliano, L.; Salmi, A.; Atzeni, E. Modelling energy source and powder properties for the development of a thermal FE model of the EBM additive manufacturing process. *Addit. Manuf.* **2017**, *14*, 49–59. [CrossRef]
16. Galati, M.; Snis, A.; Iuliano, L. Experimental validation of a numerical thermal model of the EBM process for Ti6Al4V. *Comput. Math. Appl.* **2018**, *78*, 2417–2427. [CrossRef]
17. Qi, H.B.; Yan, Y.N.; Lin, F.; Zhang, R.J. Scanning method of filling lines in electron beam selective melting. *Proc. Inst. Mech. Eng. Part B J. Eng. Manuf.* **2007**, *221*, 1685–1694. [CrossRef]
18. Zäh, M.F.; Lutzmann, S. Modelling and simulation of electron beam melting. *Prod. Eng.* **2010**, *4*, 15–23. [CrossRef]
19. Shen, N.G.; Chou, K. Thermal Modeling of Electron Beam Additive Manufacturing Process Powder Sintering Effects. In Proceedings of the Asme International Manufacturing Science and Engineering Conference, Notre Dame, IN, USA, 4–8 June 2012; American Society of Mechanical Engineers: New York, NY, USA, 2012; pp. 287–295.
20. Tolochko, N.K.; Arshinov, M.K.; Gusarov, A.V.; Titov, V.I.; Laoui, T.; Froyen, L. Mechanisms of selective laser sintering and heat transfer in Ti powder. *Rapid Prototyp. J.* **2003**, *9*, 314–326. [CrossRef]
21. Sih, S.S.; Barlow, J.W. Emissivity of powder beds. In *Proceedings of the Solid Freeform Fabrication Symposium, Austin, TX, USA, 1995*; The University of Texas at Austin: Austin, TX, USA; Volume 6, pp. 7–9.
22. Cheng, B.; Price, S.; Lydon, J.; Cooper, K.; Chou, K. On Process Temperature in Powder-Bed Electron Beam Additive Manufacturing: Model Development and Validation. *J. Manuf. Sci. Eng. ASME* **2014**, *136*, 1–12. [CrossRef]
23. Galati, M.; Snis, A.; Iuliano, L. Powder bed properties modelling and 3D thermo-mechanical simulation of the additive manufacturing Electron Beam Melting process. *Addit. Manuf.* **2019**, *30*, 100897. [CrossRef]
24. Jamshidinia, M.; Kong, F.; Kovacevic, R. Temperature distribution and fluid flow modeling of electron beam melting® (EBM). In Proceedings of the ASME International Mechanical Engineering Congress and Exposition, Proceedings (IMECE), Houston, TX, USA, 9–15 November 2012.
25. Jamshidinia, M.; Kong, F.; Kovacevic, R. The Coupled CFD-FEM Model of Electron Beam Melting® (EBM). In Proceedings of the ASME District F-Early Career Technical Conference Proceedings ASME District F-Early Career Technical Conference, ASME District F–ECTC 2013, Birmingham, AL, USA, 2–3 November 2013.
26. Vastola, G.; Zhang, G.; Pei, Q.X.; Zhang, Y.-W. Controlling of residual stress in additive manufacturing of Ti6Al4V by finite element modeling. *Addit. Manuf.* **2016**, *12*, 231–239. [CrossRef]
27. Shen, N.; Chou, K. Simulations of Thermo-Mechanical Characteristics in Electron Beam Additive Manufacturing. In Proceedings of the ASME 2012 International Mechanical Engineering Congress and Exposition, Houston, TX, USA, 9–15 November 2012; American Society of Mechanical Engineers: New York, NY, USA, 2012; pp. 67–74.
28. Cheng, B.; Chou, K. Geometric consideration of support structures in part overhang fabrications by electron beam additive manufacturing. *Comput. Des.* **2015**, *69*, 102–111. [CrossRef]
29. Jamshidinia, M.; Kong, F.; Kovacevic, R. Numerical modeling of heat distribution in the electron beam melting® of Ti-6Al-4V. *J. Manuf. Sci. Eng.* **2013**, *135*, 61010. [CrossRef]
30. Yan, W.; Liu, W.K.; Lin, F. An effective Finite Element heat transfer model for Electron Beam Melting An effective Finite Element heat transfer model for Electron Beam Melting process. In Proceedings of the Advances in Materials and Processing Technologies Conference, Madrid, Spain, 14–18 December 2015.
31. Yan, W.; Ge, W.; Smith, J.; Lin, S.; Kafka, O.L.; Lin, F.; Liu, W.K. Multi-scale modeling of electron beam melting of functionally graded materials. *Acta Mater.* **2016**, *115*, 403–412. [CrossRef]
32. Olanipekun, A.T.; Abioye, A.A.; Emmanuel, K. Time-Dependent Ginzburg—Landau equation modelling of electron beam additive manufactured Titanium alloy. *Leonardo Electron. J. Pract. Technol.* **2018**, *32*, 93–102.
33. Michaleris, P. Modeling metal deposition in heat transfer analyses of additive manufacturing processes. *Finite Elem. Anal. Des.* **2014**, *86*, 51–60. [CrossRef]
34. Galati, M.; Minetola, P.; Rizza, G. Surface Roughness Characterisation and Analysis of the Electron Beam Melting (EBM) Process. *Materials* **2019**, *12*, 2211. [CrossRef] [PubMed]

35. Shrestha, S.; Chou, K. A build surface study of Powder-Bed Electron Beam Additive Manufacturing by 3D thermo-fluid simulation and white-light interferometry. *Int. J. Mach. Tools Manuf.* **2017**, *121*, 37–49. [CrossRef]
36. Ek, R.K.; Rännar, L.E.; Bäckstöm, M.; Carlsson, P. The effect of EBM process parameters upon surface roughness. *Rapid Prototyp. J.* **2016**, *22*, 495–503.
37. Koike, M.; Martinez, K.; Guo, L.; Chahine, G.; Kovacevic, R.; Okabe, T. Evaluation of titanium alloy fabricated using electron beam melting system for dental applications. *J. Mater. Process. Technol.* **2011**, *211*, 1400–1408. [CrossRef]
38. Piscopo, G.; Atzeni, E.; Salmi, A. A hybrid modeling of the physics-driven evolution of material addition and track generation in laser powder directed energy deposition. *Materials* **2019**, *12*, 2819. [CrossRef]
39. SIMULIA User Assistance 2019—Progressive Element Activation. Available online: https://help.3ds.com/2019/English/DSSIMULIA_Established/SIMACAEANLRefMap/simaanl-c-elemactivation.htm?ContextScope=all (accessed on 9 April 2020).
40. Cagran, C.; Wilthan, B.; Pottlacher, G.; Roebuck, B.; Wickins, M.; Harding, R.A. Thermophysical properties of a Ti–44%Al–8%Nb–1%B alloy in the solid and molten states. *Intermetallics* **2003**, *11*, 1327–1334. [CrossRef]
41. Galati, M.; Di Mauro, O. Numerical prediction of surface roughness of as-built electron beam melting parts: Experimental data 2020. Available online: https://data.mendeley.com/datasets/cgwvch6kn8/1 (accessed on 16 June 2020).
42. Triantaphyllou, A.; Giusca, C.L.; Macaulay, G.D.; Roerig, F.; Hoebel, M.; Leach, R.K.; Tomita, B.; Milne, K.A. Surface texture measurement for additive manufacturing. *Surf. Topogr. Metrol. Prop.* **2015**, *3*, 024002. [CrossRef]
43. Wang, P.; Sin, W.J.; Nai, M.L.S.; Wei, J. Effects of processing parameters on surface roughness of additive manufactured Ti-6Al-4V via electron beam melting. *Materials* **2017**, *10*, 1121. [CrossRef] [PubMed]
44. Gong, H.; Rafi, K.; Karthik, N.V.; Starr, T.; Stucker, B. Defect morphology in Ti–6Al–4V parts fabricated by selective laser melting and electron beam melting. In Proceedings of the 24th Annual International Solid Freeform Fabrication Symposium—An Additive Manufacturing Conference, Austin, TX, USA, 12–14 August 2013; pp. 12–14.
45. Singh, S. *Handbook of Mechanical Engineering*, 2nd ed; S. Chand Publishing: New Delhi, India, 2011.

crystals

MDPI

Article

Microstructure and Mechanical Properties of Tempered Ausrolled Nanobainite Steel

Jing Zhao [1], Dezheng Liu [1,*], Yan Li [1], Yongsheng Yang [2,*], Tiansheng Wang [3] and Qian Zhou [4]

1 School of Mechanical Engineering, Hubei Key Laboratory of Power System Design and Test for Electrical Vehicle, Hubei University of Arts and Science, Xiangyang 441053, China; zhaojing@hbuas.edu.cn (J.Z.); liyan@hbuas.edu.cn (Y.L.)
2 School of Chemistry and Chemical Engineering, Hubei Key Laboratory of Biomass Fibers and Eco-dyeing and Finishing, Wuhan Textile University, Wuhan 430073, China
3 State Key Laboratory of Metastable Materials Science and Technology, Yanshan University, Qinhuangdao 066004, China; tswang@ysu.edu.cn
4 College of Materials Science and Engineering, Hebei University of Engineering, Handan 056038, China; zhouqian@hebeu.edu.cn
* Correspondence: liudezheng@hbuas.edu.cn (D.L.); ysyang@wtu.edu.cn (Y.Y.)

Received: 8 June 2020; Accepted: 1 July 2020; Published: 3 July 2020

Abstract: The microstructures and mechanical properties of ausrolled nanobainite steel, after being tempered at temperatures in the range of 200–400 °C, were investigated in this study. After being tempered, bainitic ferrite is coarsened and the volume fraction of retained austenite is reduced. The hardness and ultimate tensile strength decrease sharply. The impact energy, yield strength, and elongation increase with elevated tempered temperature at 200–300 °C but decrease with elevated tempered temperature when the samples are tempered at 350 °C and 400 °C. The fracture appearance of all the samples after impact tests is a brittle fracture. The variation of the mechanical properties may be due to partial recovery and recrystallization.

Keywords: tempered; ausrolled nanobainite; microstructures; mechanical properties

1. Introduction

Recently, due to their excellent mechanical properties and extraordinarily slender ferrite plates, high-C carbide-free nanostructured bainite steels have attracted more and more attention [1–6]. They are widely used in railway frogs, bearings, and automobile sheets in many industries [7–9]. Compared to martensite of similar composition, the hardness is relatively lacking in sensitivity to quite severe tempering [1]. For bainitic microstructures, the ferrite is prevented from coarsening because of the intense precipitation of carbides, thus defending the hardness and strength [10]. However, the retained austenite of nanobainite can decompose upon tempering at elevated temperatures. Bhadeshia et al. [11,12] reported that the toughness decreased because of the retained austenite decomposing and the carbides precipitating. The reason is that austenite can enhance the ductility by means of the TRIP effect. During the tensile process, the plasticity on account of martensitic transformation aids in delaying the initiation of plastic instabilities [13–15].

Caballero et al. [16] investigated the distribution change of alloying elements in a novel nanocrystalline steel during tempering. They found that during tempering, retained austenite decomposes and can reach the interface before full equilibrium. Furthermore, via a para-equilibrium transformation mechanism, cementite precipitates from supersaturated ferrite. Hasan et al. [17] investigated the tensile test of tempered carbide-free nanostructured bainitic steel. Their results suggest that, after tempering, the ductility is lowered at equal strength. The work hardening capacity is lowered due to the loss of retained austenite. Moreover, after tempering, the bainite plates coarsened slightly.

The dislocation density reduced after tempering, thus, the strength decreased. Kang et al. [18] reported the effect of tempering at 240–450 °C on the microstructure and mechanical properties of bainitic steel. By maintaining a certain degree of plasticity, the samples tempered at 340 °C reached an optimum match of strength and toughness. When the temperature was lower than 360 °C, the microstructure was insensitive to tempering. After tempering at 450 °C, the bainite plates coarsened and fine cementites were precipitated.

The weldability of high-C carbide-free nanostructured bainite steels is poor; therefore, by ausforming, medium-C nanostructured bainite is prepared. Ausforming can depress the M_S value and thus the austempered temperature decreases, and so nanostructured bainite in medium-C steel is produced [19–24].

In the aforementioned studies, the influence of tempering on the microstructures and mechanical properties of unausrolled bainite was researched. However, very few reports exist on the influence of tempering on ausrolled bainite. Therefore, the effect of tempering on the microstructures and mechanical properties of ausrolled nanobainite was researched.

2. Materials and Methods

In the study, the chemical composition of the experimental steel is presented in Table 1. Samples ausrolled at 500 °C with a reduction of 30% and austempered at 200 °C for 15 h in Reference [19] were each tempered at 200–400 °C at an interal temperature of 50 °C for 1 h. The microstructures of the ausrolled–austempered and ausrolled–austempered–tempered samples were examined by optical microscopy (OM, Axiover 200MAT, Zeiss, Heidenheim, Germany) and transmission electron microscopy (TEM, JEM-2010, JEOL, Musashino, Japan). The samples were cut along the longitudinal section with wire arc cutting. Electron-backscattered diffraction microscopy (EBSD) was used to characterize the morphology and orientation. Scanning electron microscopy (SEM, Hitachi S–3400 and Hitachi S–4800, Hitachi, Tokyo, Japan) was used to examine the morphology of the impact fracture surface. The volume fractions of retained austenite in ausrolled–austempered and ausrolled–austempered–tempered samples were determined by X-ray diffraction. The OM and XRD specimens were mechanically ground with waterproof abrasive paper. Then the specimens were polished and finally etched with Nital solution. The (TEM, JEM-2010, JEOL, Musashino, Japan) specimens were cut into ~0.6-mm thicknesses with wire arc cutting. Then the samples were mechanically ground down to ~30 μm using waterproof abrasive paper. Finally the samples of ~30 μm were thinned to perforation on a TenuPol-5 twin-jet electropolishing device (TenuPol-5, Stael, Denmark) at ambient temperature and 40 V. The electrolyte composed of 7% perchloric acid and 93% glacial acetic acid. XRD θ-2θ step scanning was performed to detect the fractions of retained austenite. The step width was 0.02° and the counting time was 2 s. Tensile properties were measured on an MTS universal material testing machine. The tensile samples had a gauge length of 25 mm along the rolling direction. Two samples are tested for each process. The hardometer (FM-ARS 9000, Tokoyo, Japan) was used to determine the Vickers hardness. The load was 1 kgf. The impact toughness was elevated with an impingement pendulum impact testing machine at room temperature. The Charpy impact samples had dimensions of 5 mm × 10 mm × 55 mm and a V-shaped notch.

Table 1. Chemical composition of the experimental steel (wt.%).

C	Mn	Si	Ni	Mo	P	S
0.540	1.890	1.700	1.630	0.240	0.005	0.002

3. Results and Discussion

The optical micrographs of the untempered and tempered samples are shown in Figure 1. The black or dark grey needles are bainite. The white blocks are retained austenite. It can be seen that the microstructures of the ausrolled–austempered–tempered samples have no obvious change compared with the ausrolled–austempered samples in Reference [19] (Figure 1a). In addition, the tempered temperature has little effect on the optical micrographs due to the low magnification. Therefore, for further investigation, the microstructures of the samples were examined by TEM.

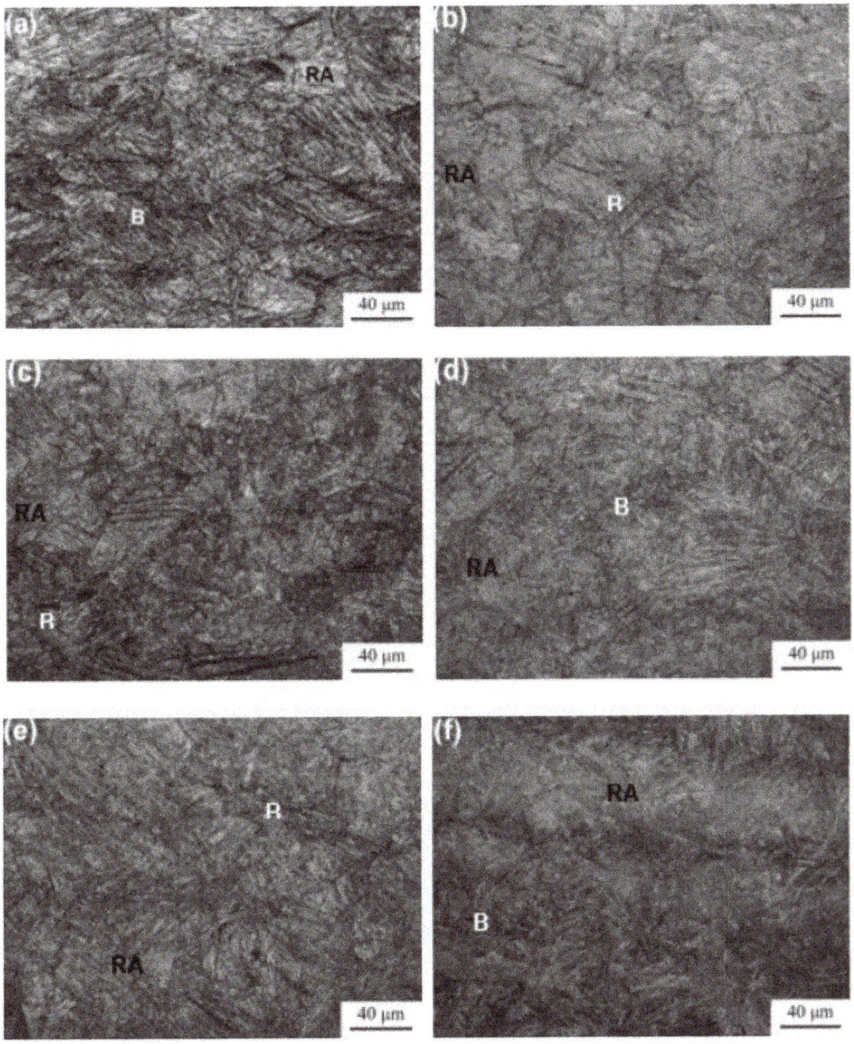

Figure 1. Optical microscopy (OM) of the samples untempered and tempered at different temperatures for 1 h following being isothermally transformed at 200 °C after ausrolling at 500 °C: (**a**) untempered; (**b**) 200 °C; (**c**) 250 °C; (**d**) 300 °C; (**e**) 350 °C; (**f**) 400 °C.

Figure 2 shows typical TEM micrographs of the untempered and tempered samples. The bright laths were bainitic ferrite and the dark laths between the bainite plates were retained austenite. The mean thickness (t) of the laths was stereologically corrected in terms of $L_T = \pi\ t\ /2$ [25]. The mean lineal intercept L_T was evaluated by measuring in a direction normal to the lath length. The mean thickness of the bainitic laths for the tempered samples were confirmed to be 144 ± 17 nm to 168 ± 35 nm, as presented in Figure 3. Compared with the untempered samples in Reference [19] (Figure 2a), the bainite ferrite coarsened after tempering.

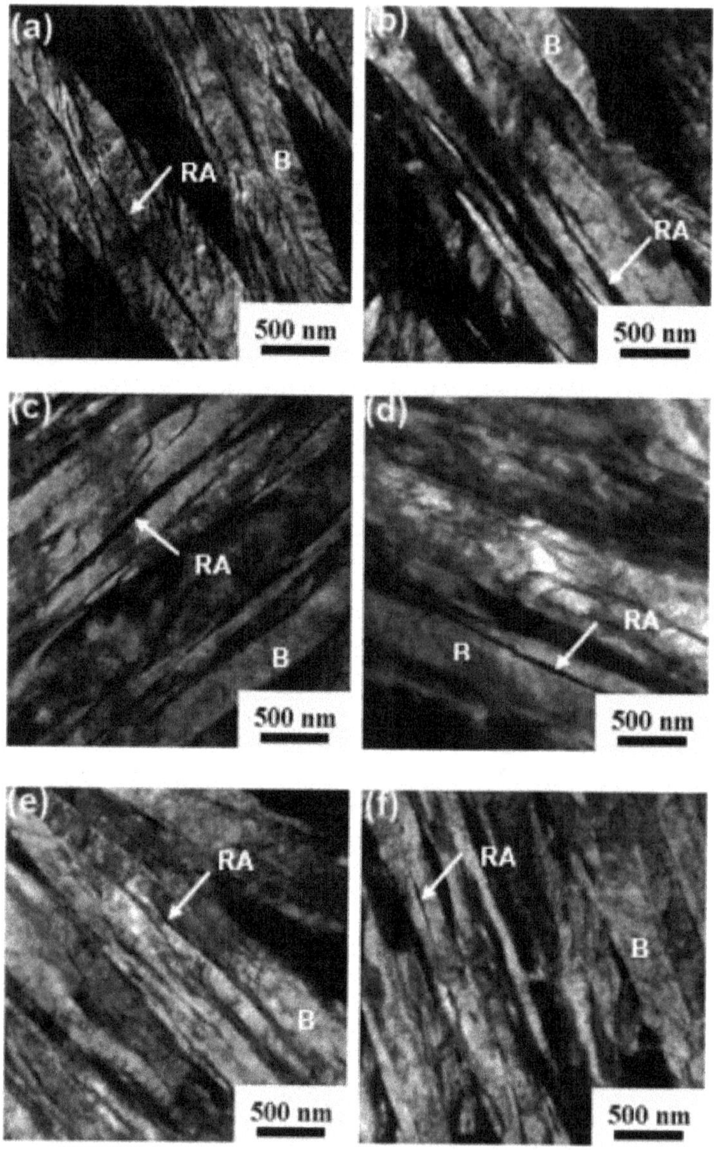

Figure 2. TEM of the samples untempered and tempered at different temperatures for 1 h following being isothermally transformed at 200 °C after ausrolling at 500 °C: (**a**) untempered; (**b**) 200 °C; (**c**) 250 °C; (**d**) 300 °C; (**e**) 350 °C; (**f**) 400 °C.

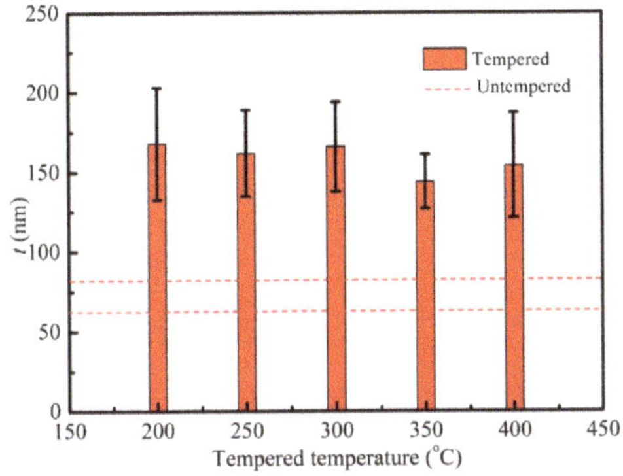

Figure 3. Bainitic ferrite lath thicknesses of untempered and tempered samples at different temperatures.

From Figure 3, it can be seen that when the tempered temperature is at 200–300 °C, there is not much difference in the thickness of bainite ferrite. However, the thicknesses of bainite ferrite are reduced when the tempered temperature reaches 350 °C and 400 °C, the reason for which may be the precipitation of carbides occurring at lath boundaries, which prevents the ferrite from coarsening.

Figure 4 presents the XRD patterns and dependence of retained austenite volume fraction on tempered temperature of tempered samples. In the XRD patterns, the peaks of ferrite (α) and retained austenite (γ) are present. Using a direct comparison method, the volume fractions of the retained austenite were determined by XRD [26]. There were 16.3%, 15.9%, 15.1%, 14.9%, and 14.8% retained austenite in samples tempered at 200–400 °C. The volume fractions of the retained austenite are arranged in order of the elevated tempered temperatures. The volume fractions of the retained austenite slightly decreased with the elevated tempered temperatures.

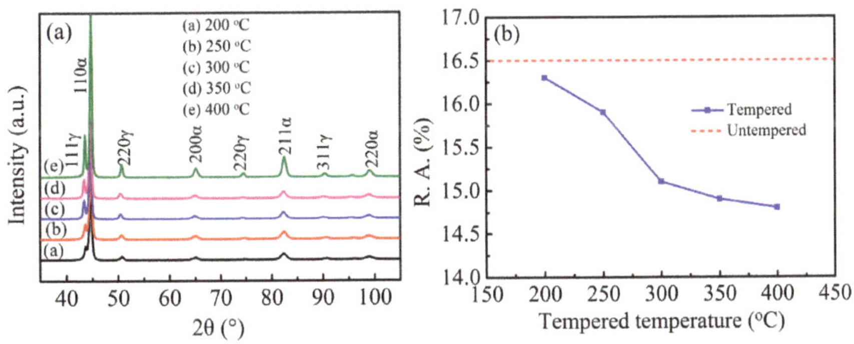

Figure 4. XRD patterns and dependence of retained austenite volume fraction on tempered temperature of samples isothermally transformed at 200 °C after ausrolling at 500 °C: (**a**) XRD patterns and (**b**) dependence of retained austenite volume fraction on tempered temperature.

Figure 5 presents the EBSD microstructure characterization of the ausrolled and tempered samples. From Figure 5, it can be seen that the high grain-boundary misorientation angle occupies a great part among the bainitic sheaves and the number fraction of the high grain-boundary misorientation angle increased with elevated tempered temperatures. However, the samples tempered at 200–300 °C show the maximum value of number fraction at 0–5°. When the tempered temperatures reached 300 °C, the number fraction of the low grain-boundary misorientation angle decreased and that of the high grain-boundary misorientation angle increased. The peaks occurred at 43° when the samples were tempered at 300–400 °C.

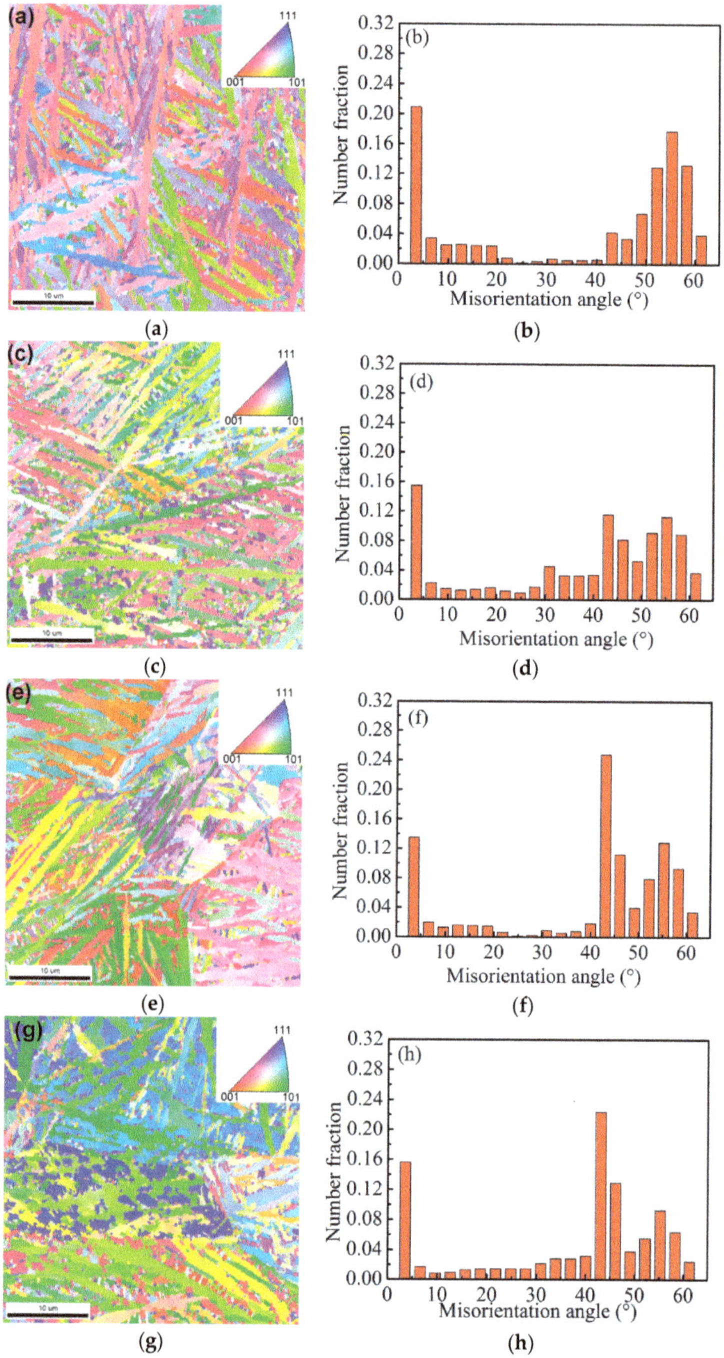

Figure 5. *Cont.*

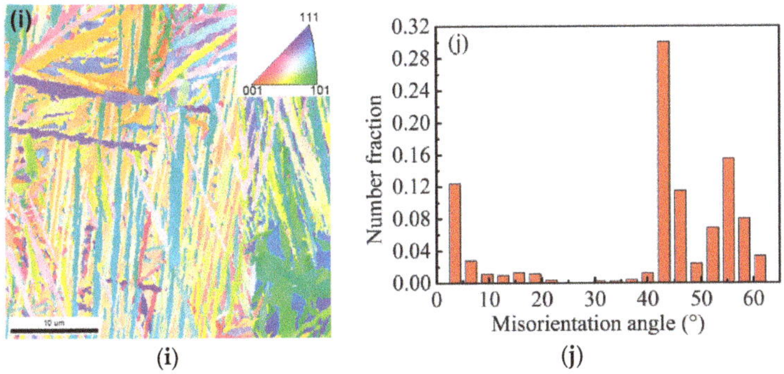

Figure 5. Electron-backscattered diffraction microscopy (EBSD) microstructure characterization of samples isothermally transformed at 200 °C after ausrolling at 500 °C and tempering at different temperatures: Orientation image (left) and misorientation-angle distribution (right): (**a**,**b**) 200 °C; (**c**,**d**) 250 °C; (**e**,**f**) 300 °C; (**g**,**h**) 350 °C; (**i**,**j**) 400 °C.

Kang et al. [18] investigated the EBSD microstructure characterization of samples isothermally transformed at 270 °C and tempered at 240, 320, and 450 °C. The results indicated that the 240 and 320 °C samples exhibited the maximum value of number fraction at 40–50° and 55° misorientation angles. However, when the tempered temperature reached 450 °C, the number fraction of the 55° misorientation angles remained unchanged and the proportion of 40°–50° misorientation angles decreased obviously. The result of this study is different from that of Kang et al. [18], which may be caused by the deformation of supercooled austenite.

Figure 6 presents the dependence of hardness and impact energy of the samples on tempered temperature. From Figure 6a, it can be seen that the hardness decreased after tempering and with elevated tempered temperature. The hardness decreased slowly during tempering at 200 °C and 300 °C, and then decreased sharply during tempering at 300–400 °C. Figure 6b illuminates the impact of energy increases after tempering, but it is not a monotonous increase. The impact energy increased during tempering at 200 °C and 300 °C, and then decreased during tempering at 300–400 °C.

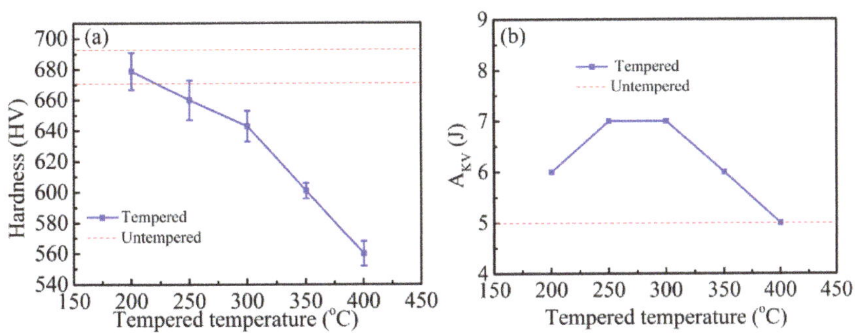

Figure 6. Dependence of hardness and impact energy on the tempered temperature of samples isothermally transformed at 200 °C after ausrolling at 500 °C: (**a**) hardness and (**b**) impact energy.

Figure 7 presents the dependence of ultimate tensile strength (UTS), yield strength (YS), and elongation (El) on tempered temperature of the samples. UTS decreased after tempering and with elevated tempered temperature. YS increased after tempering and first increased and then decreased with elevated tempered temperature. El increased after tempering, except for the samples tempered at 200 °C.

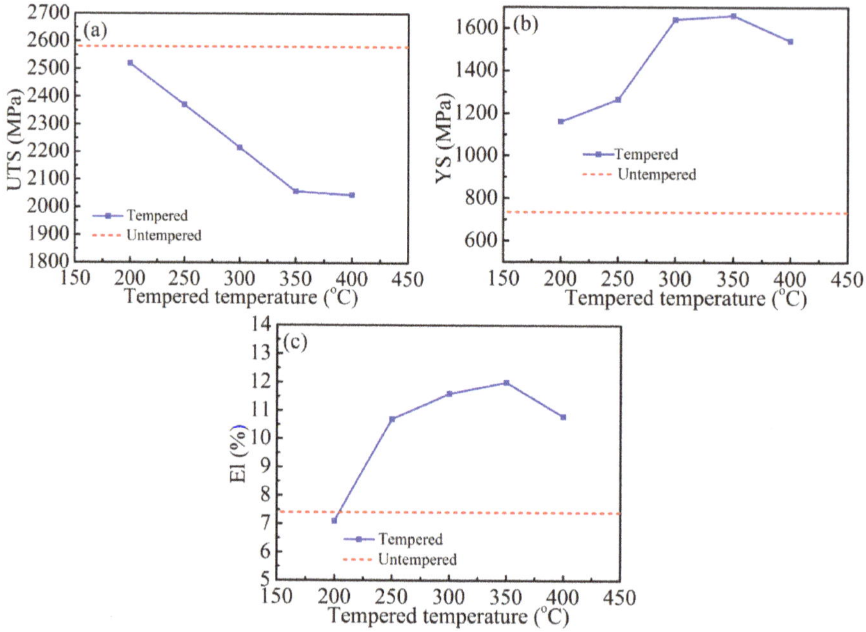

Figure 7. Dependence of ultimate tensile strength (UTS), yield strength (YS), and elongation (El) on tempered temperature of samples isothermally transformed at 200 °C after ausrolling at 500 °C: (**a**) UTS, (**b**) YS, and (**c**) El.

Supercooled austenite was reduced by 30% at 500 °C and then quickly immerged in a salt bath and isothermally quenched at a temperature of 200 °C for 15 h. Partial recovery and recrystallization did not occur due to the low temperature. Therefore, the hardness and UTS decreased slightly due to the coarsened bainitic ferrite when the tempered temperature was 200 °C. When the tempered temperature exceeded 200 °C, the microstructure of ausrolled bainite may exhibit partial recovery and recrystallization during the tempering. During the recovery, the dislocation density decreased. Therefore, the hardness and UTS decreased with increasing tempered temperature. The prior austenite grain was refined during recrystallization and YS, El, and impact energy were increased with elevated tempered temperature.

Figure 8 presents the fracture morphologies of samples tempered at different temperatures for 1 h following being isothermally transformed at 200 °C after ausrolling at 500 °C after impact testing. The fracture appearance of all the samples is brittle fracture, and a river pattern was observed in the samples. Almost no tearing ride can be seen in the samples tempered at 200–300 °C, but large quantities of tearing rides are obviously observed in the samples tempered at 350 °C and 400 °C. This phenomenon is in accordance with the difference in impact results.

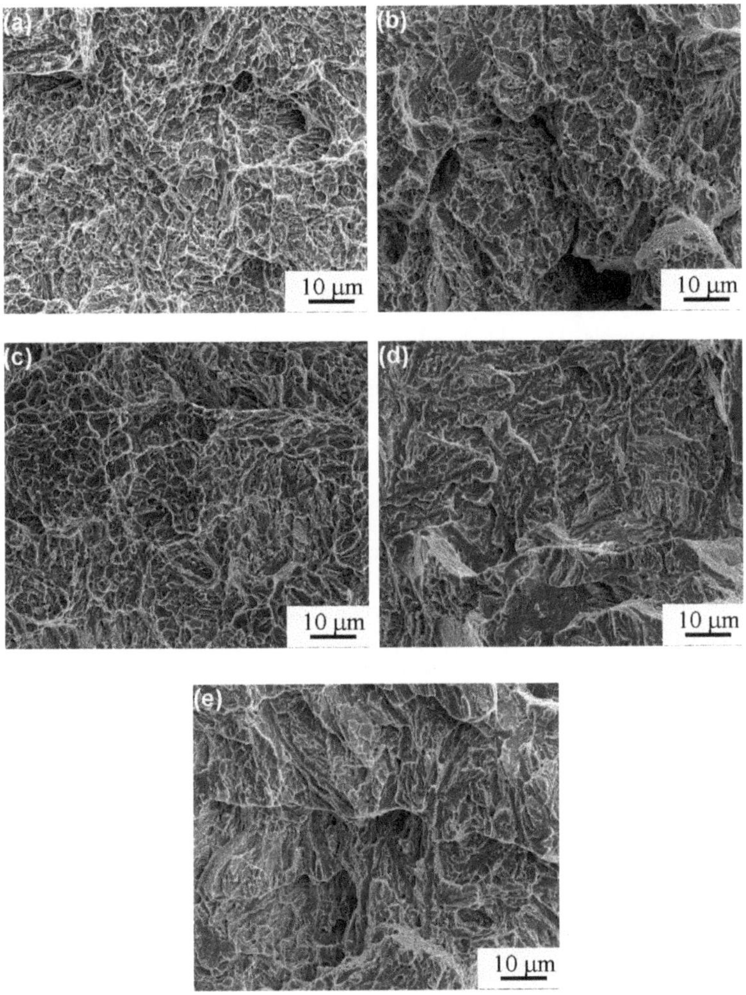

Figure 8. Fracture morphologies of samples tempered at different temperatures for 1 h following being isothermally transformed at 200 °C after ausrolling at 500 °C after impact testing: (**a**) 200 °C; (**b**) 250 °C; (**c**) 300 °C; (**d**) 350 °C; (**e**) 400 °C.

4. Conclusions

The effect of tempering on microstructures and mechanical properties of ausrolled–austempered samples was investigated. In summary, after tempering, the bainite ferrite coarsened from 70 ± 9 nm to 144 ± 17 nm to 168 ± 35 nm. The volume fraction of retained austenite was reduced, and the impact energy was slightly enhanced. The hardness and UTS decreased sharply. The hardness decreased from 682 ± 11 $HV_{1.0}$ to 679 ± 12 to 560 ± 8 $HV_{1.0}$. The UTS decreased from 2581 to 2520–2045 MPa. The YS and El increased, but not monotonously. The impact energy, YS, and El increased with elevated tempered temperature when the samples were tempered at 200–300 °C, but decreased with elevated tempered temperature when they were tempered at 350 °C and 400 °C. When the tempered temperature exceeded 200 °C, the microstructure of ausrolled bainite may have exhibited partial recovery and recrystallization during tempering, so the hardness and UTS decreased, and the YS, El, and impact energy were increased with increasing tempered temperature. The fracture appearance of all the samples after impact testing is brittle fracture.

Author Contributions: Conceptualization, T.W. and D.L.; writing—original draft, J.Z.; methodology, Y.L. and Q.Z.; writing—review and editing, D.L. and Y.Y. All authors have read and agreed to the published version of the manuscript.

Funding: This work was supported by Hubei Superior and Distinctive Discipline Group of "Mechatronics and Automobiles" (No. XKQ2020002), Hubei Key Laboratory of Power System Design and Test for Electrical Vehicle and the National Natural Science Foundation of China (Grant No. 51801051).

Conflicts of Interest: The authors declare no conflict of interest.

References

1. Caballer, F.G.; Bhadeshia, H.K.D.H.; Mawella, K.J.A.; Jones, D.G.; Brown, P. Very strong low temperature bainite. *Mater. Sci. Technol.* **2002**, *18*, 279–284. [CrossRef]
2. Garcia-Mateo, C.; Caballero, F.G.; Bhadeshia, H.K.D.H. Development of hard bainite. *Isij Int.* **2003**, *43*, 1238–1243. [CrossRef]
3. Caballero, F.G.; Bhadeshia, H.K.D.H. Very strong bainite. *Curr. Opin. Solid State Mater. Sci.* **2004**, *8*, 251–257. [CrossRef]
4. Caballero, F.G.; Miller, M.K.; Babu, S.S.; Garcia-Mateo, C. Atomic scale observations of bainite transformation in a high carbon high silicon steel. *Acta Mater.* **2007**, *55*, 381–390. [CrossRef]
5. Caballero, F.G.; Miller, M.K.; Garcia-Mateo, C. Carbon supersaturation of ferrite in a nanocrystalline bainitic steel. *Acta Mater.* **2010**, *58*, 2338–2343. [CrossRef]
6. Wang, B.Q.; Song, X.Y.; Peng, H.F. Design of a spheroidization processing for ultrahigh carbon steels containing Al. *Mater. Des.* **2007**, *28*, 562–568. [CrossRef]
7. Zhang, F.C.; Yang, Z.N.; Kang, J. Progress in bainitic steel used for railway crossing. *J. Yanshan Univ.* **2013**, *37*, 1–7.
8. Sswley, K.; Sun, J. Advanced rail steels:investigating the bainitic option. *Railw. Track Struct.* **1997**, *3*, 22–27.
9. Zhang, F.C.; Wang, T.S.; Zhang, P.; Zheng, C.L.; Lv, B.; Zhang, M.; Zheng, Y.Z. Sliding wear and low cycle fatigue properties of new carbide free bainitic rail steel. *Scr. Mater.* **2008**, *59*, 294–296. [CrossRef]
10. Peet, M. *Transformation and Tempering of Low-Temperature Bainite*; University of Cambridge: Cambridge, UK, 2010.
11. Bhadeshia, H.K.D.H.; Edmonds, D.V. Bainite in Silicon Steels: New Composition–property Approach Part1. *Metal. Sci.* **1983**, *17*, 411–419. [CrossRef]
12. Bhadeshia, H.K.D.H.; Edmonds, D.V. Bainite in Silicon Steels: New Composition–property Approach Part2. *Metal. Sci.* **1983**, *17*, 420–425. [CrossRef]
13. Sherif, M.Y.; Garcia-Mateo, C.; Sourmail, T.; Bhadeshia HK, D.H. Stablility of retained austenite in trip-assisted steels. *Mater. Sci. Technol.* **2004**, *20*, 319–322. [CrossRef]
14. Sherif, M.Y. *Characterisation and Development of Nanostructured, Ultrahigh Strength, and Ductile Bainitic Steels*; University of Cambridge: Cambridge, UK, 2005.
15. Bhadeshia, H.K.D.H. Nanostructured bainite. *Proc. R. Soc. A.* **2010**, *466*, 3–18. [CrossRef]
16. Caballero, F.G.; Miller, M.K.; Garcia-Mateo, C.; Capdevila, C.; Babu, S.S. Redistribution of alloying elements during tempering of a nanocrystalline steel. *Acta Mater.* **2008**, *56*, 188–199. [CrossRef]
17. Hasan, H.S.; Peet, M.J.; Avettand-Fenoel, M.N.; Bhadeshia, H.K.D.H. Effect of tempering upon the tensile properties of nanostructured bainitic steel. *Mater. Sci. Eng. A.* **2014**, *615*, 340–347. [CrossRef]
18. Kang, J.; Zhang, F.C.; Yang, X.W.; Lv, B.; Wu, K.M. Effect of tempering on the microstructure and mechanical properties of a medium carbon bainitic steel. *Mater. Sci. Eng. A.* **2017**, *686*, 150–159. [CrossRef]
19. Zhao, J.; Guo, K.; He, Y.M.; Wang, Y.F.; Wang, T.S. Extremely high strength achievement in medium-c nanobainite steel. *Scr. Mater.* **2018**, *152*, 20–23. [CrossRef]
20. Wang, T.S.; Zhang, M.; Wang, Y.H.; Yang, J.; Zhang, F.C. Martensitic transformation behaviour of deformed supercooled austenite. *Scr. Mater.* **2013**, *68*, 162–165. [CrossRef]
21. Zhang, M.; Wang, Y.H.; Zheng, C.L.; Zhang, F.C.; Wang, T.S. Austenite deformation behavior and the effect of ausforming process on martensite starting temperature and ausformed martensite microstructure in medium-carbon si–al-rich alloy steel. *Mater. Sci. Eng. A.* **2014**, *596*, 9–14. [CrossRef]
22. Zhang, M.; Wang, T.S.; Wang, Y.H.; Yang, J.; Zhang, F.C. Preparation of nanostructured bainite in medium-carbon alloysteel. *Mater. Sci. Eng. A.* **2013**, *568*, 123–126. [CrossRef]

23. Zhang, M.; Wang, Y.H.; Zheng, C.L.; Zhang, F.C.; Wang, T.S. Effects of ausforming on isothermal bainite transformation behaviour and microstructural refinement in medium-carbon si–al-rich alloy steel. *Mater. Des.* **2014**, *62*, 158–174. [CrossRef]
24. Zhao, J.; Jia, X.; Guo, K.; Jia, N.N.; Wang, Y.F.; Wang, Y.H.; Wang, T.S. Transformation behavior and microstructure feature of large strain ausformed low-temperature bainite in a medium C-Si rich alloy steel. *Mater. Sci. Eng. A.* **2017**, *682*, 527–534. [CrossRef]
25. Chang, L.C.; Bhadeshia, H.K.D.H. Austenite Films in Bainitic Microstructures. *Mater. Sci. Technol.* **1995**, *11*, 874–881. [CrossRef]
26. De, A.K.; Murdock, D.C.; Mataya, M.C.; Speer, J.G.; Matlock, D.K. Quantitative Measurement of Deformation-induced Martensite in 304 Stainless Steel by X-ray Diffraction. *Scr. Mater.* **2004**, *50*, 1445–1449. [CrossRef]

MDPI
St. Alban-Anlage 66
4052 Basel
Switzerland
Tel. +41 61 683 77 34
Fax +41 61 302 89 18
www.mdpi.com

Crystals Editorial Office
E-mail: crystals@mdpi.com
www.mdpi.com/journal/crystals

www.ingramcontent.com/pod-product-compliance
Lightning Source LLC
Chambersburg PA
CBHW062042220926
51850CB00010B/2635
* 9 7 8 3 0 3 9 4 3 1 4 0 3 *